PRINCIPLES & APPLICATIONS OF
SOLAR ENERGY

PAUL N. CHEREMISINOFF, P. E.
Associate Professor
Environmental Engineering
New Jersey Institute of Technology
Newark, New Jersey

THOMAS C. REGINO
Energy Resources Group
Pandullo Quirk Associates
New York, New York

ANN ARBOR SCIENCE
PUBLISHERS INC./THE BUTTERWORTH GROUP

ENERGY TECHNOLOGY SERIES

**GASOHOL FOR ENERGY PRODUCTION
WOOD FOR ENERGY PRODUCTION
BIOGAS PRODUCTION AND UTILIZATION
FUNDAMENTALS OF WIND ENERGY
PRINCIPLES AND APPLICATIONS OF SOLAR
 ENERGY
GASOHOL SOURCEBOOK**

Additional volumes will cover coal, hydrogen, hydro-electric and geothermal generation, ocean thermal energy conversion, and insulation.

Third Printing, 1981
Second Printing, 1979

Library of Congress Catalog Card No. 78-50308
ISBN 0-250-40247-5

Butterworths, Ltd., Borough Green, Sevenoaks, Kent TN15 8PH, England

FOREWORD

The idea of using solar energy is not new. Man has benefited from, employed and even worshipped this energy in different ways for thousands of years. The sun, a colossal furnace, pours out tremendous energy, of which we on earth receive a small portion. Potential forms of energy dependent on the sun's power include the oceans, the wind, hydropower, fossil fuels and biomass.

The *solar age* now beginning promises to slow our dependence on increasingly expensive, scarce and environmentally questionable fossil fuels and nuclear energy. As a group, the solar energy technologies are by no means the greatest scientific challenge ever faced. Principles and material requirements intrinsically appear more manageable than those associated with large-scale nuclear reactions. Many of the solar conversion methodologies described in this volume are best suited to decentralized operation. The authors feel this is one of solar energy's main strengths as well as potential weaknesses.

The authors have attempted to present this overview work in an easy-to-read form. The goal has been to provide a summary which, hopefully, will save the reader time and shelf space. The book's chapters represent the major aspects of solar energy technology. The authors have exercised their best efforts to distill the data from a large number of reliable sources. The reader may want to further evaluate information, so primary references and citations have been given wherever possible. Finally, our thanks also to the many individuals and organizations in this field who gave of their time, expertise and knowledge in filling in the informational gaps and reviewing the material presented.

Paul N. Cheremisinoff

Thomas C. Regino

iii

Paul N. Cheremisinoff, P.E., is Associate Professor of Environmental Engineering at New Jersey Institute of Technology, a consulting engineer, Associate Editor of *Water & Sewage Works*, and editor of two newsletters. He is author-editor of several major engineering handbooks; Diplomate of the American Academy of Environmental Engineers, Fellow of the New York Academy of Sciences, Sigma Xi.

Thomas C. Regino is a graduate of Newark College of Engineering and received his M.S. in Environmental Engineering from New Jersey Institute of Technology. His work in environmental and mechanical engineering has involved him in numerous projects dealing with sewage treatment, solid wastes management, solar energy systems and energy resources. He is presently affiliated with the Energy Resources Group of Pandullo Quirk Associates.

CONTENTS

1. SOLAR ENERGY—HISTORICAL 1
   Introduction 1
   B.C. 1
   A.D. 2
   The Seventeenth Century 2
   The Eighteenth Century 3
   The Nineteenth Century 5
   The Twentieth Century 9

2. SOLAR ENERGY AVAILABILITY 17
   Introduction 17
   Sun to Earth 17
   From Space to Your Door 18
   Solar Position 21
   Insolation 24
   Incident Angle 26

3. THERMAL COLLECTION DEVICES 31
   Nonconcentrating Collectors—Flat-Plate Type 33
   Liquid-Cooled Flat-Plate Collectors 34
   Air-Cooled Flat-Plate Collectors 44
   Nonconcentrating Collectors—Evacuated Tube Type . . . 48
   Concentrating Collectors 51

4. THERMAL SOLAR ENERGY APPLICATIONS 57
   Introduction 57
   Heating of Buildings 58
   Air Systems 60
   Hydronic Systems 61
   Controls 63
   Cooling of Buildings 64
   Domestic Water Heating 67
   Solar Irrigation 69
   Thermal Generation of Electricity 71

5. PHOTOVOLTAIC GENERATION OF ELECTRICITY . . .   81
   Introduction   81
   Photovoltaic Effect   82
   Photovoltaic Devices   84
   Array Systems for Electrical Generation   92

6. ENERGY FROM THE WIND   101
   Introduction   101
   Horizontal-Axis Machines   103
   The Smith-Putnam Wind Turbine   105
   The New Approaches   108
   The ERDA-NASA Model Zero   109
   Vertical-Axis Machines   113
   Other Ideas   118
   Storage and Conversion for Use   120

7. OCEAN THERMAL GRADIENT POWER   123
   Introduction   123
   OTEC Power Applications   135
   Electricity by Cable Transmission   135
   Aluminum Production   136
   Ammonia Production   137
   Mariculture Effects   139

8. CHEMICAL CONVERSION OF SOLAR ENERGY   141
   Introduction   141
   Photoelectrolysis   142
   Electrochemical Photovoltaic Cells   150
   Hydrogen Utilization   152

9. BIOLOGICAL CONVERSION OF SOLAR ENERGY . . .   157
   Introduction   157
   Production of Photosynthetic Biomass   159
   Solid and Animal Wastes as Biomass   164
   Conversion of Biomass to Energy   165

10. A LOOK AT THE FUTURE   171

APPENDIX A   175
APPENDIX B   211
APPENDIX C   215
REFERENCES   235
INDEX   243

CHAPTER 1

SOLAR ENERGY—HISTORICAL

INTRODUCTION

Man has employed solar energy in different ways for literally thousands of years, dating back to the period before Christ. Mention has been found in surviving references of solar energy applications both esoteric and practical.

B.C.

During the fifteenth century B.C., the Egyptian ruler Amenkotep III supposedly possessed "sounding statues" that operated when the air in their base pedestals expanded after exposure to sunlight. His son, Zari Memnon, allegedly owned an artificial bird that would "sing" when sunlight fell on it.[1] The historian, Plutarch, professed that vestal virgins employed cone-shaped metal objects to light ritual fires during the seventh and eighth centuries B.C.[2] It has been recorded elsewhere that it was not uncommon, during this period in history, to use the sun's rays to distill liquids and dry agricultural products.[3]

The most dramatic of early applications, that of the Greek genius Archimedes using solar energy to defend the harbor of Syracuse against the Roman fleet in 212 B.C., is shrouded in controversy to this day. According to Greek physician and historian, Galen (130-220 A.D.), Archimedes employed a "burning mirror" to set fire to the ships of the invading Roman fleet while they were still a bowshot's distance from shore. Galen's reference to this event is found in his *De Temperamentis*. The debate that ensued over whether this feat was possible was fueled by the fact that two earlier historians, Livy (59 B.C.-17 A.D.) and Plutarch (46-120 A.D.), although having written earlier of Archimedes, made no direct reference to to the incident.[1,2] Archimedes supposedly wrote a book entitled *On*

1

Burning Mirrors, but no copies survived to settle the matter. The question remained as a source of debate and inspiration.

A.D.

Whether the Romans experienced the power of solar energy at Syracuse, it seems that they soon were beginning to recognize its uses. In 77 A.D. the chronicler Pliny wrote of the Roman practice of using burning glasses to light fires and cauterize wounds.[1]   No further reference to solar energy applications is found in the literature until more than 1,100 years later.

During the 12th century, Ioanne Zonaras wrote of the siege of Constantinople, and that the attacking fleet of Bitellius was burned at a distance by means of a large number of mirrors. A Greek named Proclus was the man to whom this Archimeidan feat was credited. It can safely be assumed that this report was heir to as much controversy as was Galen's Syracuse account of some 1000 years earlier.

At the risk of making an obvious observation, whether due to the coincidences inherent to incomplete records, the possible biases of a progression of historians or inspiration derived from an actual event, the story of Archimedes' triumph at Syracuse (and the many attempts at repeating the technique) runs like a strong thread throughout the early history of solar energy development. Whether true or not, the story is significant.

THE SEVENTEENTH CENTURY

In 1615 Salomon de Caux constructed the first device in the category of a "solar engine." His device, made of glass lenses, a supporting frame and an airtight metal vessel containing water and air, simply produced a small water fountain when the air heated up during operation. It was more a toy than a workhorse device, but is the first published account of solar energy application since the fall of the Roman Empire.

A physicist, Athanasius Kircher (1601-1680) performed experiments during the mid-1600s with mirrors, attempting to set fire to wood at a distance—another effort to rediscover the then 1,800-year-old technique of Archimedes. Perhaps because constructing numerous small-scale mirrors is simpler than large-scale lens fabrication partially accounts for the repeated interest in burning mirrors. Another point in favor of mirror-type experimentation for larger-scale applications could have been the simpler principles involved in planar reflection as opposed to the optics of lenses (refraction).

A mathematician from Germany, Ehrenfried Walter von Tschirnhaus (1651-1708), worked with lens-type concentrators. A member of the

French Academy of Science, von Tschirnhaus designed and constructed lenses up to some 30 in. in diameter (larger than, but not as precise as, any telescope lens then in existence). Using these lenses, he was able to melt ceramic materials with concentrated sunlight. The Duke of Orleans later obtained one of the von Tschirnhaus lenses and made it available for use by his physician, Homberg, who used it to melt gold and silver for various purposes.

In 1695 two Italian empiricists, Targioni and Averani, used a concentrating mirror during their experiments. The goal was to melt a diamond with solar energy. The outcome was not recorded, but is of little consequence historically. Their use of concentrated solar energy to extend man's knowledge of the limits and properties of materials is a tried and true experimental application, however, and is significant.

This century was indicative of a new interest in science, or rather a more favorable political climate in which it could grow. The 1600s produced the momentum that made the Industrial Revolution possible, as well as the progress in technology (solar energy included) that went with it.

THE EIGHTEENTH CENTURY

A French scientist, George Louis Leclerc Buffon (1707-1788), was responsible for a series of multiple-mirror solar furnaces, the largest consisting of some 360 small planar mirrors, individually positioned to focus on a common point. (France still is among the leaders in solar furnace technology. Her megawatt furnace at Odeillo-Fond-Romeau is the second largest such device in the world.[4]) In the French Royal Gardens, in 1747, using a furnace with 168 mirrors, Buffon ignited a woodpile from a distance of about 195 ft. He concluded that Archimedes could have set those Roman ships on fire, but only from a distance of 100-150 ft.[1] This conclusion was almost certainly based on the assumption that the ancient Greek was privy to a much less sophisticated technology 1,900 years earlier, and this could not have equaled Buffon's device. The validity of this judgment remains to be determined. D. L. Simms wrote recently that "there are ample historical, scientific and military grounds for concluding that Archimedes did not use a burning mirror as a weapon of war."[5] Simms feels that Archimedes would have been unable to calculate the radiant flux necessary to do the job, and would not have used a method with such a low probability of success. The question remains.

Claude Poillet, a countryman of Buffon's and a mathematician-scientist of the time, was one of the first to be concerned with measuring sunlight intensity. He estimated that "the usable energy from the sun per square yard of the earth's surface between the equator and approximately 43° N or S latitude was about 1/6 thermal unit per second, which corresponds to nearly one horsepower."[2] Poillet's area of interest, measuring sunlight intensity and relating it to geographical position, has proved to be at least as important in solar energy utilization as the hardware itself.

Jacques Cassini, a French astronomer working with the Paris Observatory, built a concentrating mirror about 45 in. in diameter, which came to be known as the "Royal Mirror." In 1747 the mirror was presented to King Louis XV. The mirror, according to Observatory records, achieved temperatures above 1800° F, and thus was able to melt silver to such a low viscosity that it formed "spidery filaments when plunged into cold water."[2]

In 1744 an English scientist named Joseph Priestly used a burning glass to heat some mercuric oxide and collected the gas produced during the process. This constituted the historic first deliberate synthesis of oxygen, one of the experiments from which Priestly concluded that air is not an elementary substance, but a mixture of oxygen and other gases.[6] He went on to discover that oxygen was also a waste product of plants, a piece of information important to later work on evolution.

A French contemporary of Priestly's, Antoine Lavoisier, also ran the mercuric oxide experiment, backwards as well as forwards. In addition, he measured both the oxygen production and uptake of the process. The source of heat used by Lavoisier was sunlight concentrated through a burning glass. He was able to augment the importance of Priestly's discovery of oxygen by incorporating it into his theory that combustion, rather than creating a substance then known as phlogiston, was merely a process of oxidation. Lavoisiers' combustion theory is fundamental to modern chemistry, and was made possible by Priestly's findings.

The solar furnace that Lavoisier designed, built and used in his work was also worthy of mention. It consisted of a hollow double convex glass vessel filled with alcohol, forming a liquid lens. In conjunction with a smaller solid lens, this furnace achieved temperatures as high as 3200° F. It is quite doubtful whether solar energy was essential to the research of these two men, but Lavoisier did observe that "the fire of ordinary furnaces seems less pure than that of the sun."[6] The fact that solar heat is composed of pure energy and contributes no trace contamination to a specimen is probably its prime advantage as far as materials research is concerned.

Nicholas de Saussure (1740-1799), a Swiss naturalist, conducted the first recorded experiments in applying the sun's rays for cooking purposes. He called his solar oven a "heat box," and it was composed of multiple separated glass covers positioned over a blackened surface. The bottom and sides of the oven were surrounded with insulation, and eventually it

was recorded that the oven achieved a temperature of 320° F. De Saussure did, indeed, use the solar oven to prepare food.[2]

THE NINETEENTH CENTURY

A slightly different solar oven was introduced in 1837 by an astronomer
from England, John Fredrick Herschel. The son of German-born astronomer
Sir William Herschel, J. F. built a small solar oven while on a trip to
Africa's Cape of Good Hope. He constructed it of mahogany, painted it
black and buried it in the sand for purposes of insulation. A double-glazed
cover, the only portion of the device left exposed, served to minimize heat
losses through the top, while letting in sunlight. The oven registered a
maximum temperature of about 240° F, and was used throughout the expedition by Herschel and his staff to cook both meat and vegetables.[1,2]

The famous Sir Henry Bessemer (1813-1898) was also connected with
solar energy experimentation. The originator of the Bessemer process of
steel manufacturing, Sir Henry designed and fabricated a solar furnace some
10 ft in diameter, incorporating about 100 small planar mirror elements
oriented towards a common focus. Using the furnace, Bessemer successfully melted copper and zinc. It seems, however, that his interest in experimenting with alloys of carbon and iron (steel) was greater than his concern
for solar energy, since he soon lost interest in the device.[2]

Several papers were published in the 1860s by C. L. A. Callier, a French
physicist. Although no experimental work by Callier is described, the
papers "demonstrate considerable knowledge of the subject."[2]

One of those truly deserving of the label "solar energy pioneer" is
Augustin Mouchot (1825-1911), a physics professor from France. He initiated 20 years of government-funded experimentation with reflector-type
concentrators in 1860. His efforts led him to conceive, design and build
the first reflector devices based on the shape of a truncated cone (Figure
1-1). These devices, now known as axicons, were intended to focus sunlight along the central axis of the cone, rather than to a single-point focus.
This arrangement allows the radiation to be uniformly distributed along
the outside of a tubular energy-absorbing surface, thus alleviating some of
the danger of burning a hole in an underfilled boiler (inherent in point-
focus devices).

Mouchot took full advantage of this configuration, and built several
solar-powered steam engines between 1864 and 1882. His reflectors were
built of "silver-plated sheet-metal plates, suitably mounted so that the entire device could be easily turned to follow the position of the sun. The
collecting surface (of a typical Mouchot assembly)had an area of 40 ft[2]
and was connected to a boiler which received about 87% of the sun's

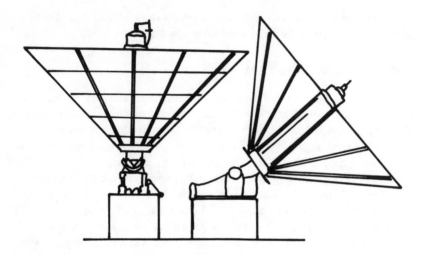

Figure 1-1. A typical Mouchot axicon solar concentrator. The device focused solar energy, in a line focus, along the tubular absorber in the device's center.

heat."[2] The device was connected, unfortunately, to a steam engine of extremely low efficiency, yielding a poor overall performance.

Working together with another Frenchman, Abel Pifre, Mouchot built an axicon-type solar reflector which was put on exhibition in 1882 at the Tuileries Garden in Paris. It supplied energy to a steam engine which simultaneously drove a printing press. A complementary newspaper was printed by this device, having the pertinent title *Le Soleil,* which was distributed to passers-by.

In 1879 Mouchot published a book entitled *La Chaleur Solaire et ses Applications Industrielles,* which included descriptions of his experimental efforts. The book also contained the first "scientific survey of the economic potential of solar energy...."[2] The survey concluded that large-scale applications of solar energy were not economically feasible at that time. It is quite possible, however, that Mouchot's work may have helped to spur additional interest in solar energy research and development by contemporary and subsequent scientists.

John Ericsson (1803-1889), a Swedish-American famous for creating the armor-clad ship *Monitor* during the American Civil War, devoted his time between 1868 and 1886 to solar energy utilization. He built both paraboloid and parabolic cylindrical reflecting surfaces, which he used to concentrate sunlight onto blackened tubular boilers (Figure 1-2). These devices produced enough steam to power a 2.5 horsepower (hp) engine, and one of them was exhibited at several industrial fairs throughout New York State.

Figure 1-2. The parabolic cylinder concentrator designed by Ericsson. This device is a forerunner of the parabolic trough concentrators used today to power solar irrigation equipment.

Ericsson expended about $90,000 of his personal funds during his experimentation. Perhaps it was this fact or Mouchot's precedent that prompted him to consider the economic feasibility of solar-driven engines. Nevertheless, he reached the conclusion that solar-powered devices were an order of magnitude (10 times) more costly than conventional power sources capable of performing the same work. Ericsson reasoned further that the extra expense was warranted only for solar energy applications in remote areas of the planet that receive large amounts of solar radiation. Unfortunately, although this inventor seems to have had a more comprehensive and sophisticated grasp of the necessary principles than did most of his contemporaries, none of his work on solar energy was published.

The first large-scale solar distillation operation was built in northern Chile in 1872. It was designed by Charles Wilson, from England, and covered 51,000 ft^2 of ground surface. Its function was to extract freshwater from saltwater, for consumption at a nearby nitrate-mining installation. Shallow troughs filled with saltwater were constructed and covered with slanting glass plates. The glass, performing its now familiar function of allowing sunlight to enter while trapping heat energy (for absorption by the saltwater), allowed that water to vaporize. Air was passed between the glass covers and the saltwater troughs, cooling the water vapor, and allowing it to condense on the underside of the glass plates. From there, the now freshwater would trickle down the slanting glass and drop into channels, to be collected, ready for use. This operation was a very successful low-technology application that performed well for 40 years, until the nitrate mine was closed down. Its maximum production amounted to 6,000 gpd of freshwater.

Also active in the field during the 19th century were two Germans—Stock and Heynemann. They experimented with a solar furnace composed of several glass "plano-convex lenses with a diameter of about 30 in. and a focal length of 20 in."[2] These experiments were the first to be concerned with the provision of heat within a vacuum enclosure. Sunlight was concentrated through the lenses to a focal point coincident with the specimen to be heated. The specimen was held in a magnesia crucible, which, in turn, was placed within a highly evacuated glass vessel. The procedure proved that it was possible to melt samples of silicon, copper, iron and manganese with sunlight in the absence of air.

In 1878 an Englishman, William Adams, published a small book concerning solar engines. At the time, Adams was serving in Bombay, India. His most powerful prototype allowed for continuous daytime operation of a 2-kW water pump near his bungalow at Middle Colaba, in Bombay.[1] Adams was interested in experimenting with solar cookers as well. His best device consisted of glass planar mirrors arranged in the shape of an inverted eight-sided pyramid.[2] Adams found that he could, indeed, cook food with the device, but found the taste more repulsive than appealing.

An American scientist and aeronaut, Samuel Pierpont Langley, was yet another individual interested in solar ovens. In 1884 Langley used an insulated box to cook food during an expedition to Mount Whitney in California. The increased solar intensity at those high altitudes apparently was sufficient to overcome the lower air temperature and pressure.[2]

In the 1880s E. Weston proposed using a thermocouple device for the generation of electricity. Solar energy was focused on a "hot junction" made up of a pair of wires composed of dissimilar metals. A "cold junction" of an identical pair of wires was to be kept at or below ambient temperatures. The temperature differential created between the hot and cold junctions resulted in a voltage differential as well, and electric current flowed. This seems to have been the first thermoelectric system for solar energy utilization. Patents were issued to H. C. Reagan, Jr. in 1897, H. F. Cottle in 1898 and R. H. Dunn in 1899 for other thermoelectric schemes.

A patent was issued to M. L. Severy in 1893 for his concept of using banks of wet storage batteries as a means of storing the excess electricity generated during the day by a solar engine. While it is highly unlikely that this specific system was ever built, the concept lives on of storage as a way to attenuate solar energy usefulness and increase its feasibility. Severy generated alternative versions of his concept up until 1909.

In 1896 a different patent was issued to C. G. O. Barr for another type of solar engine. Barr's idea was a big one, including an array of semi-parabolic mirrors "mounted on railroad cars on a circular track, with a fixed boiler at the focus of the system."[1] This concept was essentially a simple form of a subsequent Soviet design of decades later, and probably also served as embryo to the United States' "Power Tower" concept.[1] While Barr's system and the Soviet system were abandoned, the U.S. power tower system is still on the drawing board. (This will be discussed in more detail in Chapter 4.)

THE TWENTIETH CENTURY

While the 1800s saw more solar research take place than at any other time, the 20th century brought solar energy technology to new levels of diversification. While compiling a complete list of solar energy research and experimentation during the first half of this century would be a full-time job, the following text attempts to document major developments and trends.

Patents were issued to E. H. McHenry in 1900 and 1911 for subatmospheric-pressure solar engines that used two working fluids to produce power. Water was used to collect heat, and "a working fluid of lower boiling point than water to drive the engine."[1] E. C. Ketcham received a patent for a similar system in 1905. The two-fluid system was introduced to avoid the low efficiency problems that had been observed by Augustin Mouchot and others.

In Pasadena, California, in 1901, a large axicon-type concentration device was built by A. G. Eneas. The structure had 642 ft^2 of solar collection surface, developed some 4.25 hp, had a claimed 74.6% collection efficiency, and operated at a working steam pressure of 150 lb/in^2.[8] This system was unique at that time, in that it was the first large solar power device to operate with "simultaneous double movement."[9] This movement, compensating the mirror orientation for both daily and seasonal solar position changes, was obtained with a hanging weight, clock-type mechanism. Eneas built a similar, slightly larger, axicon in 1903 in Mesa, Arizona, and another in Willcox, Arizona, in 1904.

The first experiments with flat-plate collectors began in 1902, carried out by H. E. Willsie and John Boyle, Jr. As will be described further in Chapter 3, the flat-plate device was a very shallow box with a black internal surface, a clear glass cover plate, and was cooled by some form of transfer fluid flow, usually water. Although the flat-plate collector does not concentrate sunlight as do lens-type and mirror-type collectors, it is able to utilize diffuse as well as direct radiation and thus is able to operate during more widely varying weather conditions. Willsie and Boyle's collector used the heated water to vaporize some volatile liquid (such as ammonia, sulfur dioxide, ether, etc.). The performance of their collector, although it was built of admittedly crude materials, was such that "even in cold, raw October weather (temperatures) were high enough to vaporize sulfur dioxide for the engine."[1]   The first system was built near Olney, Illinois, during 1902.

Subsequent work with two-fluid engines encouraged the formation of the Willsie Sun Power Company, which built an ammonia-driven solar engine system in St. Louis, Missouri, in 1904. It incorporated some 600 ft^2 of glass collector area, and drove a 6-hp engine, with an 85% collection efficiency.[9]   The company's two largest (and last) engines were constructed in Needles, California by Boyle and Edward Wyman, in 1905. They were fitted with sulfur dioxide engines of 15 and 20 hp, respectively, and operated irrigation equipment bordering the Colorado River. The technical performance of the Willsie systems was up to expectations, but they simply could not compete economically with the conventional steam-type installation of the period, being some two to four times as costly. The company was dissolved, but not until after having proved the technical feasibility of solar equipment, as well as demonstrating that flat-plate collectors also have a place in the solar scheme of things.

In 1904 a Portuguese priest, Father Himalaya, was responsible for the design and fabrication of a large parabolic-horn solar furnace exhibited at the St. Louis World's Fair (Figure 1-3). The furnace stood approximately 42 ft high, and provided a concentration ratio of about 2,000:1. Hundreds of 2 in. x 4 in. planar mirrors were used to cover the mirrored surface and, as always, the main problems to be overcome were size and monetary limitations.

Flat-plate collector research was begun in 1907 by Frank Shuman, of Philadelphia. Water, as usual, was used as the heat transfer fluid. Like Willsie and Boyle, Shuman's system used a second fluid to drive a steam engine; in this case, ether (Figure 1-4). His installation was located in Tacony, Pennsylvania, and included 1200 ft^2 of collection surface. The engine developed 3.5 hp with a boiler temperature of up to 240° F. Shuman founded the Eastern Sun Power Company Limited in 1908, encouraged by the technical performance of his units.

A large, 10,300 ft^2 system of flat collectors was built at Tacony in 1911. Twenty-six banks of collectors incorporated rows of mirrors to

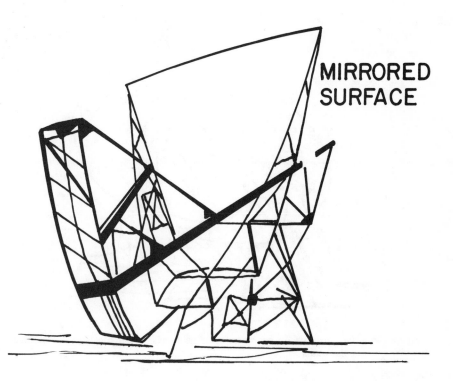

MIRRORED SURFACE

Figure 1-3. The Himalaya solar furnace of 1904. Exhibiting a concentration factor of 2000, the device was able to smelt metals.

boost performance, doubling the effective solar input to the system. This was a one-fluid system that vaporized water into low-pressure steam which, in turn, drove a reciprocating pump used to deliver water against a head of 33 ft. Shuman estimated that the installation would have generated 100 hp, had it been located in a tropical environment. The approximate cost of the system was $200/hp.[9]   Sun Power was gearing up to market similar systems, but a bigger, more important challenge soon presented itself to Shuman and his associates.

In collaboration with a Professor C. V. Boys in 1912, Shuman began design and construction of one of the most technologically significant solar power installations yet attempted, a 50-hp solar engine project at Meadi, Egypt (Figure 1-5). The goal was to use sunlight to power irrigation equipment along the Nile River.[3]   Solar energy was concentrated in

Figure 1-4. The Eneas axicon concentrator. The tubular boiler can be seen mounted in the center.

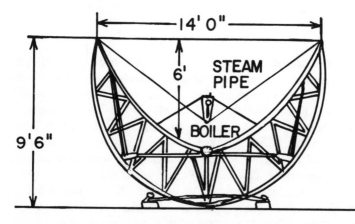

Figure 1-5. The solar irrigation project designed by Shuman and Boys at Meadi, Egypt. A 100-hp solar engine was driven by the collector field at about 50% of capacity.

cylindrical troughs, whose inner surface was covered with small planar mirrors. These cylinders rotated about their north-south axis to track the sun throughout each day's operation. The cross-sectional shape of the troughs approximated a parabolic curve. This arrangement focused sunlight on a line, rather than a point, similar to Mouchot's axicon, but with a more uniform temperature distribution. A solar interception surface of over 13,000 ft[2], with a concentration ratio of 4.5:1,[8] was provided using seven 205-ft-long collectors.[9] The heat-absorbing boilers were placed at the focal-line, and the steam generated within them served to feed the 100-hp engine onsite. This engine "developed between 50 and 60 horsepower continuously on one five-hour run."[9] Economic inefficiency forced its abandonment during World War I.[1] The team of Shuman and Boys was granted a patent on their engine in 1917.

A solar thermoelectric device was built in 1908 in Russia by W. Zerassky. The hot and cold junctions were made up of zinc-antimony and silver-plated wires. One of the two wire pairs was placed in a small solar over-like box to form the "hot" junction. Various other schemes on this thermoelectric theme were patented and discussed during the next decade.

J. A. Harrington stored solar power by hydraulic means some 60 years ago in New Mexico. During the day, a solar engine was used to pump water into a 5,000-gal tank 20 ft high. At night the water was released to drive a turbine/dynamo for electric lighting. This was "one of the earliest attempts to store solar-generated power."[8] Patents were issued on various solar engine ideas during the 1920s to W. J. Harvey, L. H. Shipman and Robert H. Goddard, the pioneer of rocket technology. Unfortunately, no working prototypes based on those ideas were built. An operating engine was built in Italy in 1929 by Cesare Romagnoli, but using water and ethyl chloride as the heat transfer and driving fluids. His engine was used for irrigation purposes.[2]

Probably the most significant event to occur in the 1920s in solar energy was the initiation of a lifetime of work by Dr. Charles Greeley Abbot, a man connected with the Smithsonian Institution for the vast majority of his life. Abbot published his first work on solar energy in 1926, although he first studied the sun some 21 years earlier, at the age of 33. A parabolic trough collector of Abbot's that ran a 0.5-hp engine was exhibited at the International Power Conference in Washington, DC, in 1936. Abbot also "introduced an ingenious solar flash boiler which permits rapid development of steam within five minutes after solar exposure."[9] A 0.20-hp capacity version of this device was used in Florida in 1938.

Dr. Abbot instituted a long program of solar observation at the Smithsonian Institution in Washington. The program monitored solar activity and radiation levels in a scientifically sound manner over a period of many years. His data has served as a basis for much of the subsequent research concerned with the sun. He published his last paper in 1973, at the respectable age of 100.

The late 1920s and 1930s saw a considerable market for solar water heaters develop in this country—a response to the economic pressure created by the Depression to exploit "free" energy sources. Most of these devices were centered in southern Florida, with its plentiful year-round sunshine.[10] Late in the 1930s, however, cheap natural gas became available. This event, coupled with general economic recovery, wiped out most of the market for solar heaters in America. The few surviving companies have recently experienced an unprecedented surge in business, however, and are gearing up for production once again.[10] At present, there are some 25,000 units, installed in Florida in the 1920s, that are still working—effective testimony to their cost-effectiveness. The 1930s saw the introduction of solar hot water heating in Japan as well. The market there has been slow but steady. As early as 1960, it was estimated that there were over 200,000 heaters in use.[10]

Thus far, we have seen solar energy applied to distillation, generation of electricity, solar furnaces, solar engines, irrigation, cooking applications and hot water heating. In the late 1930s the time was ripe for someone to explore solar energy's worth in space heating applications. As part of the Godfrey L. Cabot Solar Energy Conversion Research Project, the first solar heated house was built at the Massachusetts Institute of Technology in 1939 under the direction of H. C. Hottel and B. B. Woertz. The structure "was a two-room laboratory building intended primarily for use in developing the methods of calculating collector performance, and for that purpose was provided with a water storage tank large enough to collect heat in the summer and store it for the winter."[9]

The "house" had no auxiliary heating system, but storage temperatures were sufficient throughout the year to demonstrate technical success. Economically, the experiment showed that long-term storage was much too costly to be practical. The major disadvantage of the research results was that very large heat storage systems were thought of by the public as a necessary part of a workable solar space heating system.[9] This house was to become but the first in a series of four M.I.T. solar houses, a research program conducted by M.I.T. until 1962. M.I.T. Solar House I was demolished in 1946 to make way for Solar House II.[11] Although M.I.T. Solar House I and the research results from work on other solar applications appeared both promising and eventually profitable, World War II put a total halt to solar energy research in this country for the duration.

Dr. Frederico Molero, working at the Helio Power Laboratory of the Power Institute of the U.S.S.R. Academy of Sciences, was not barred from his work on solar energy, however. He used parabolic concentrating surfaces to generate steam in an installation at Tashkent, Uzbek Province, from 1941-1946. The steam was generated "for use in irrigation, livestock, watering, refrigeration, processing and heating. Full details are not available."[9]

Aside from the six-year moratorium on research imposed by the war effort, World War II generally had a "shot in the arm" effect on scientific development. Avenues explored and methods developed during the early 1940s for war-related projects furnished new tools with which peacetime research could be made more effective. The public had also seen how the tremendous advancements in manufacturing technology and productivity brought on by the war could affect their lives. These and other reasons explain the faster pace of scientific development programs in the postwar years.

The postwar boost to solar research was delayed for a few years, however. While the newer, underdeveloped nations immediately recognized the desirability of solar utilization, they possessed insufficient resources with which to develop its usefulness. The major powers, on the other hand, saw no need to exploit the sun when conventional fuels were so plentiful and cheap. As such, the solar research effort in the "advanced" nations immediately following the war was none too concerted, and produced little in the way of results.

Enthusiasm was starting to build, however, and the late 1940s and the 1950s witnessed optimistic projections as to solar energy use by the public at large.[1] Solar research programs began to grow and multiply on many fronts. Space heating research continued at M.I.T. with Solar Houses II and III. Dr. George Löf installed a collector of his own design on a house in Boulder, Colorado in 1945. In 1948, Dr. Maria Telkes and Eleanor Raymond built a solar-heated house in Dover, Massachusetts (this was the first house to store heat via eutectic salts). Raymond Bliss and Mary Donovan installed a solar heating system on a 25-year-old house owned by the U.S. Forest Service near Tucson, Arizona, in 1954.[12]

Also in the 1950s, the Phillips Research Laboratory in Eindhoven, Holland, conducted a successful large-scale research program concerned with the fabrication of hot air motors. Their engine reportedly achieved an efficiency equal to that of contemporary internal combustion engines. Dr. Telkes was also active at this time in working with solar distillation devices, having built and operated such a system in 1951 at Cohasset, Massachusetts. Two Indian scientists, Ghai and Khanna, constructed a solar energy system in New Delhi that used the heat from a parabolic collection surface to develop $1/6$ hp in hot-air engine operating at 700-1200° F. The system operated from 1950-1955. Khanna immediately built another installation some four times larger than the first, also in New Delhi.[9]

Professor Amelio of Bari, Italy, used ethyl chloride to drive turbines at a solar installation in Libya, in 1954. His system reportedly cut friction losses and was able to produce superatmospheric pressures. A countryman of his, Enzo Carlivari, also used an ethyl chloride motor on the island of Ischia. His system was said to have developed 4.5-hp and used some 8.8 lb/sec of 158° F water. The motor reportedly operated at 8,000 rpm.

Various conferences were organized in America during this period, the largest being a World Symposium on Applied Solar Energy, held in Tucson and Phoenix, Arizona, in 1955, with 130 delegates from more than 37 countries. Those attending numbered some 900 registrants, representing not only the sciences, but also "industry, finance, government and education."[9] The solar energy subtopics discussed covered a myraid of interests, including residential applications as well as those agricultural and industrial. The Symposium listed more than two dozen countries in which solar research was being conducted at that time.

The scope of research carried on during the intervening two decades defies summary. Let it suffice to say that while research did indeed increase during the 1950s and 1960s, its economic and technological scope has literally been approximating a process of explosion since the first fuel crisis in the early 1970s. The present state of the art is such that only subsets of the solar energy field can be studied in any depth. Subsequent chapters of this book will attempt to give the reader overviews of what problems each subfield of solar energy technology addresses.

CHAPTER 2

SOLAR ENERGY AVAILABILITY

INTRODUCTION

A concept to which all of us have been exposed at some time in our lives is the image of our sun as an enormous heat source, the "furnace" of our solar system. Indeed, before man came up with the concept of "solar system," he equated "sun" and "warmth." It is an obvious relationship.

In fact, aside from comparatively small contributions from gravitational and nuclear interactions, every process that has ever occurred on this plant was fueled, directly or indirectly, by energy from our sun. With this in mind, it can be seen that the sun is the "prime mover" in this neighborhood of the universe. This fact is not directly observable with only our five senses, and required intellectual conceptualization to become evident. To better understand this vital transfer of energy, some background is necessary.

SUN TO EARTH

We are speaking of enormous amounts of energy. The sun radiates, through a continuous process of thermonuclear fusion, approximately 83.3 million billion billion kilowatt hours (8.33×10^{25} kWh^3) of energy into space every day. While the earth's daily receipt of a minute fraction of this energy depends on its distance from the sun, as well as sunspot activity on the solar surface, it always amounts to very close to 4.14 million billion kilowatt hours (4.14×10^{15} kWh^3) each day. Although this is much less than one billionth of the sun's total output, earth's total share is still an inconveniently huge amount of energy; thus, a much smaller quantity, known as the "solar constant," is used to describe energy intensity. The solar constant is defined as the density of "solar radiation on a surface normal to the sun's rays beyond the earth's atmosphere at the average earth-sun distance, 92,955,888 miles (one astronomical unit).[13] While "it has been determined from analyses of radiation data that the variations in total radiation emitted by the sun are probably less than ± 1.5%..."[14] the strong variances introduced by atmospheric penetration allow the quantity "solar constant" to be treated as if it were a constant. Its present value is commonly accepted to be 429.2 Btu/hr/ft^2 of surface.

Energy from the sun traverses that astronomical unit, in slightly more than eight minutes time, in the form of electromagnetic radiation. This radiation is made up of energies that exhibit many different wavelengths, varying from oil microns (X-rays) to some 100 meters (radio waves) in length. Conveniently, fully 99% of the sun's energy lies within a much narrower wavelength range of 0.28 μ (ultraviolet rays) to 4.96 μ (infrared rays).[13] This energy distribution can be seen in Figure 2-1. Incidentally, the human eye can detect electromagnetic radiation only within the extremely limited wavelength range of about 0.4-0.7 μ. Thus, it is obvious that the human eye can sense only a very limited piece of the action transpiring around us at any given time. It has been recognized only recently that man must employ artificially created sensors to observe various phenomena that occur beyond the range of our built-in senses. Still more recent is the ability to conceive, design and fabricate these devices. The information obtained by using these instruments has already significantly and irrevocably changed our conceptions regarding the universe.

Among the new information so far obtained is the beginning of an understanding of that energy that the earth receives from the sun. Keep in mind that the 99% (0.28-4.96 μ) radiation band refers specifically to that energy reaching the outer boundaries of earth's atmosphere.

FROM SPACE TO YOUR DOOR

A myriad of new factors are introduced in examining the last leg of sunlight's journey, through the atmosphere, to the planet's surface. While the extreme outer fringes of our atmosphere, the domain of the solar constant, receives an essentially unvarying quota of solar energy, new complications force the observer on earth's surface to switch his vocabulary from exactitudes to averages and approximations.

The first in a series of interactions between our atmosphere and incoming solar radiation takes place some 12-20 miles above the earth's surface. In this region lies a protective, recently celebrated, ozone layer. Ozone is a highly reactive molecule comprised of three atoms of oxygen, and is denoted O_3. It coexists in the upper atmosphere with two other forms

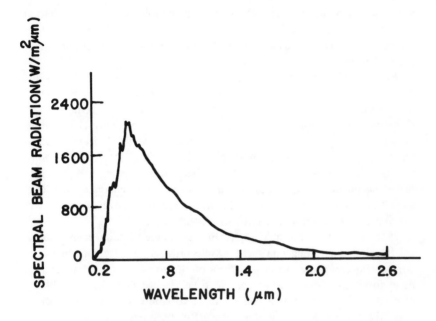

Figure 2-1. The NASA (1971) standard spectral irradiance at the mean sun-earth
distance and a solar constant of 1353 W/m^2 (429.2 Btu/ft^2/hr). The
peaked region of the curve represents the majority of sunlight's energy.
The distribution shown here is further exaggerated after passing through
the earth's atmosphere.

of oxygen: molecular oxygen, O_2 (that which we breathe) and free ele-
mented oxygen, O. The more energetic shortwavelength photons in sunlight (having wavelengths of 0.32 μ or less) are absorbed on striking O_3
and O_2 molecules, supplying the energy necessary to break the binding
forces and dissociate those molecules into O_2 and O. These simpler forms
of oxygen subsequently collide and react with other oxygen molecules
to recombine into O_3 and O_2.

The ozone layer, then, can be characterized as a steady-state condition
of dissociation and recombination. This is not to be confused with a
state of perfect equilibrium, however. The ozone concentration in our
atmosphere does vary with time and location, in response to a plethora
of forces not yet understood. The net result of the passage of solar radiation through the ozone layer is a change in the strength and character
of that radiation. The shortest wavelengths of energy that get through
in appreciable quantities are of ultraviolet radiation of greatly reduced
intensity. This filtering process is indeed fortunate, in view of the harm-
ful effects of excessive ultraviolet exposure (sunburn, skin cancer, blindness,

eventual loss of life). The vast majority of cosmic and X-type radiation are also prevented from reaching the surface by this very same absorption process, so it becomes clear that destruction of the ozone layer is an issue of tremendous import.

As the radiation further penetrates the atmosphere, gas molecules and dust particles begin to intercept their share of incoming photons. The now less-energetic radiation has insufficient strength to break most of these more stable molecules. Thus, instead of being totally absorbed, a portion of the photons undergoing collision are merely deflected, "scattered more-or-less uniformly in all directions, so that some of the radiation is redirected away from earth and back into space again."[15] This scattering phenomenon still affects the shorter, more energetic wavelengths most (including the blue end of the visible spectrum), which reveals why the sky is blue when seen from low altitudes.

Water vapor is responsible for further screening of solar energy, and chiefly affects the longer infrared wavelengths. In fact, clouds in our atmosphere can cause as much as 80% of the solar energy incident on them to bounce back into space. With an average of approximately 50% of earth's total surface area being subject to cloud cover at any given time, the magnitude of this screening effect becomes apparent.[15]

Another factor involved in sunlight's intensity at the terrestrial surface is the angle of its approach, and it is also easily understood. As can be simply visualized, light passing through the atmosphere in a direction perpendicular to the earth's surface will encounter a minimum thickness on the way down, and thus arrive at some maximum intensity (Figure 2-2). As the angle of approach deviates from perpendicularity, more and more atmosphere is encountered by the sunlight, with a corresponding decrease in intensity. This is demonstrated by relatively weak sunlight intensity near sunrise and sunset, during the winter season, or even all year around in the higher latitudes of our world. This angle of approach is called "solar altitude" and will be discussed again.

The atmosphere surrounding us is a kinetic phenomenon. Localized concentrations of its different constituents form, shift and dissipate continuously, allowing different amounts of sunlight to reach the surface at different times and locations. Predicting with precision the amount of sunlight that will definitely be incident on a specific location at a specific instant is impossible, without perfect weather forecasting (a capability beyond man's reach at present). Averages based on records of past weather behavior must, therefore, be employed to generate rough models of atmospheric effects on solar availability for any given region. In this country, the average percentage of solar constant radiation actually incident upon the terrestrial surface ranges from under 40 to over 75%.[4]

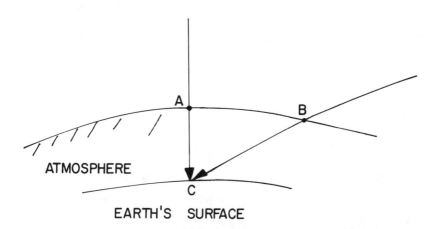

ATMOSPHERE

EARTH'S SURFACE

Figure 2-2. The effective thickness of atmosphere that sunlight must penetrate to reach the earth's surface is a crucial factor in determining its final intensity at the surface. Shown here, in amplified form, it can easily be seen that the distance from B to C is greater than that from A to C. These distances correspond to the relative amounts of filtering and scattering to which the sunlight is subjected in each case. It is precisely this phenomenon that is the main cause of seasonal weather patterns on earth.

This is a valid approach, since essentially all specific locations around the world possess their own characteristic weather patterns. A significant shift in weather patterns is necessary to change a long-term average, so reasonable accuracy can be achieved with this method.

SOLAR POSITION

One should not become overly concerned about the "haphazard" factors involved in solar availability. Superimposed on the unpredictabilities, there are strong cyclic behaviors that provide the observer with a foundation of precise regularity. These behaviors are caused by the motions of our planet, including the earth's annual revolution around the sun, the earth's daily rotation about its own axis, and the tilt of that axis with respect to the plant of earth's orbit. These distinct motions, predictably, yield characteristic effects.

In and of itself, the yearly orbit of earth causes little variance in the amount of solar radiation to which it is exposed. The extreme annual changes in the radius of earth's elliptical orbit only amount to some 2%, allowing us to think of the orbit as circular without introducing serious error. An interesting fact regarding this is that the earth is farthest from the sun during the summer months in the northern hemisphere (95.90 million miles) and closest during the winter season (89.83 million miles). A schematic diagram of this is shown in Figure 2-3. Earth's changing seasons are not caused by our orbit alone, therefore, but rather by that orbit in conjunction with another feature of earth—its "tilt."

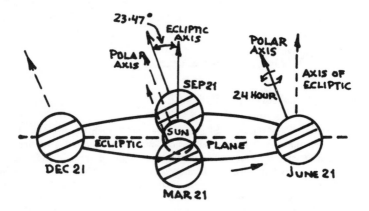

Figure 2-3. Motion of the earth about the sun.[13]

The equator of our planet is tilted some 23°, 27 minutes[13] with respect to an imaginary, but very precise, plane in space—the "ecliptic." This plane is defined by earth's orbit, and, in turn, defines the sun's equator. As we circle the sun, a slight daily variation occurs in "the angle between the earth-sun line (on the ecliptic) and the equatorial plane (of earth). This angle, called the solar declination, δ (delta), varies continuously...."[13] The variation of δ causes the earth to present a slightly different face to the sun each day, and is the motion responsible for those seasonal changes in weather we all experience each year. It also causes the mysterious annual lengthening and shortening of the intervals between sunrise and sunset, in spite of earth's very regular daily rotation about its own axis.

If you were to observe the sun's motion from any fixed position on the earth's surface for a sufficiently long period of time, the sun would be seen to exhibit regular patterns of daily movement across the sky. Of course, these patterns do vary gradually throughout the year. The position of the sun at any chosen instant can be defined fully using two angles that are measured from your fixed location. One of these angles is called the solar altitude, β (beta), and is measured vertically from the sun's apparent position in the sky to the horizon directly below it. The horizontal angle described between this point on the horizon and the direction to true south is called solar azimuth, ϕ (phi). The direction to true south must be used here, and using magnetic south instead would result in an inaccurate reading.

The values of altitude and azimuth depend upon time of year, time of day and the latitude of your location. In lieu of measuring these angles yourself, the angles can be calculated, given enough accurate time and place information. For practical purposes, however, values interpolated from the standard tables included in Appendix I will be sufficient for most applications.

As mentioned earlier, the angle along which the sunlight approaches the surface determines the effective thickness of atmosphere through which the light must pass. This significantly influences the intensity, wavelength distribution and degree of scattering undergone by incoming solar flux. Since the earth's tilt causes solar altitude to decrease to a minimum during the winter months, any increase in solar radiation received by earth, due to our planet's slightly closer position to the sun, is more than canceled out by the "thicker" atmosphere. It becomes obvious, then, that solar position is an important parameter. An understanding of it is absolutely necessary to estimate and evaluate the type and strength of solar radiation that reaches us. The knowledge afforded us by this coordinate system is of crucial importance when designing systems to capture solar energy. An accurate method of predicting solar position allows us, simply, to determine the best orientation for solar collection devices. It enables the system to utilize the maximum amount of energy available for specific geographic locations and system applications.

The explanation that follows should help lend substance to these disembodied statements. From any location in the northern hemisphere, the following sun movement is seen to occur every day: dawn occurs somewhere along the southeastern quadrant of the horizon; the sun reaches solar noon (its highest daily apparent altitude) when its position places it at some angle directly above a point on the horizon due south of the observation point; and sunset occurs somewhere in the southwestern quadrant (with an azimuth equal to and opposite that exhibited at sunrise). This behavior can be predicted quite accurately by using the solar position tables in Appendix I.

The information presented thus far shows a simple point that should be remembered when dealing with solar energy systems. Since maximum altitude allows the sunlight to pass through a minimum thickness of atmosphere, any surface oriented toward true south will receive more sunlight than if it were pointed in any other direction. For vertical surfaces (as an example), the variation in total solar radiation received with respect to orientation is shown in Figure 2-4. Having been exposed to solar position, we will now discuss sunlight's intensity.

INSOLATION

The term "solar constant" has been shown to represent the amount of solar radiation incident on a unit surface area exterior to earth's atmosphere during a unit time interval. This situation is, in fact, only a special case of the more general concept of "insolation" (as in "incoming solar radiation"). Insolation is also used to quantify incident energy per unit area per unit time, but is nearly always measured on or close to the earth's surface, giving rise to wide variations in its magnitude.

Aside from this obvious difference between the solar constant and terrestrial insolation, there are other complications. Although all "solar constant" energy consists of direct beams of solar radiation, terrestrial insolation does not. Significant portions of energy become deflected by the atmospheric scattering effects discussed previously, and thus approach the earth's surface from all positions above the horizon, not just from the direction towards the sun. This component is called "diffuse" radiation, and becomes the dominant type of solar flux on cloudy or hazy days.

Another fraction of the solar energy reaching any collection surface has just rebounded from surrounding terrain, surface water, vegetation or manmade structures in the area. This component is appropriately called "reflected" radiation. While this type of radiation is usually so small a part of the total sunlight that reaches the collection surface as to be negligible, looking at a field of snow on a bright winter day shows that this situation can change. Indeed, many systems are designed to deliberately take advantage of reflected sunlight. A general image of these three types of radiation can be seen in Figure 2-5.

Averaged over the entire United States "solar energy arrives at the surface...at an average rate of about 1500 Btu/ft^2 day."[4] While this number is so general as to be useless for any design purposes, it serves to give a rough indication of the solar energy available in this country. A little more specifically, the peak clear day insolation received in most sections of the U.S. would, predictably, occur at solar noon and reach a value of about 300 Btu /ft^2/hr. This amount of energy would, under ideal conditions, be sufficient to heat up a gallon of water about 36°F. Although ideal conditions are never achieved, the 300 Btu/ft^2/hr figure does represent the rough maximum level of insolation that can be expected on a clear day.

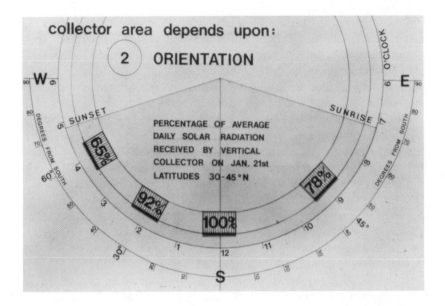

Figure 2-4. Maximum performance is achieved by facing the collectors toward true south. A variation of $20°$ or so in either direction, however, still gives good results. (Source: Sunworks, Division of Enthone, Inc.)

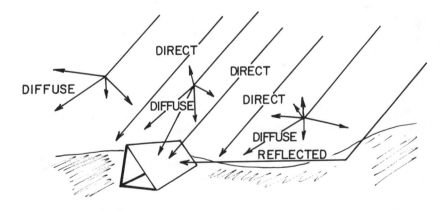

Figure 2-5. The three different components of solar radiation that strike any collection surface include direct, diffuse and reflected radiation. Although the latter two of these three types may reach significant levels with the right conditions, the direct component of sunlight is always predominant during clear weather and is the type for which solar collection systems are oriented.

It is, then, a combination of direct, diffuse and reflected radiation that strikes a collecting surface and is recorded as total insolation. This is significant when interpreting insolation data, since systems such as photovoltaic arrays (solar cells), concentrating collectors and solar furnaces can only utilize direct radiation during their operation. Subsequent chapters will show how this fact can influence system effectiveness.

Different geographical latitudes are heir to different average quantities of insolation during clear weather, as determined by the geometry of our planet and its cyclic motions with respect to the sun. The American Society of Heating, Refrigeration and Air Conditioning Engineers (ASHRAE) is presently the most widely accepted source for insolation data. Clear Day Insolation tables originally published by ASHRAE are included in Appendix I, and contain very useful information concerning expected insolation levels during clear weather at certain latitudes. As was the case with solar position data, values for specific locations can be obtained through interpolations, with little problem.

It should be remembered that clear day insolation does not, alone, give an accurate picture of available solar energy in the real world. Just as characteristic weather patterns are associated with their corresponding geographic positions, each of those areas will receive some average percentage of the total energy that would have been available given constantly clear weather. This average is called "Percent Possible Sunshine." For any given time of year, the percent possible sunshine figure multiplied by the clear day insolation (possible sunshine) figure will yield a pretty fair estimate of how much solar energy will be available at your location. Maps showing the isopercentile lines in the U.S. for each month of the year are also in Appendix I.

INCIDENT ANGLE

In most cases direct radiation is by far the predominant form of solar radiation available, and is thus the component for which solar collection systems should be designed. Direct radiation is, by definition, directional in nature. The orientation of the collecting surface with respect to that direction of incidence is extremely important to the performance of any solar system, regardless of the mechanism of collection.

Specifically, you should present the full face of your collection surface to the direct radiation coming towards it. In other words, you want the collecting surface to be perpendicular (normal) to the sun at all times, in much the same way that the imaginary surface was oriented to measure the solar constant. Systems to track the continuous motion of the sun are usually elaborate and expensive, however. For this reason, most low-technology collection systems are mounted in a stationary position. Therefore, the best course of action to take in this situation is to position the collection surface such that any difference between the normality desired and the actual situation is at a minimum. The way to achieve this minimum in the horizontal direction is to point towards true south as shown earlier in Figure 2-4. The vertical piece of the puzzle must be handled with a more general discussion.

An angle known as the "incident angle," θ (theta), is commonly used to describe collector orientation with respect to the sun. It is defined as the angle "between the direct solar ray and a line normal to the irradiated surface. The importance of the incident angle lies in the fact that it determines both the intensity of the direct radiation component striking the surface and the ability of the surface to reflect, transmit or absorb the sun's rays."[13]   Since the angle, θ, represents the difference between desired and actual, this is the quantity you'll want to minimize. To better be able to visualize this angle and the other incidental angles we are about to discuss, refer to Figure 2-6.

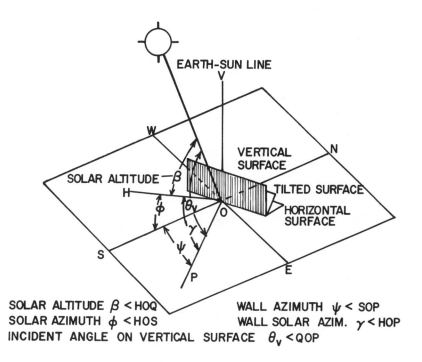

SOLAR ALTITUDE β < HOQ   WALL AZIMUTH ψ < SOP
SOLAR AZIMUTH ϕ < HOS   WALL SOLAR AZIM. γ < HOP
INCIDENT ANGLE ON VERTICAL SURFACE θ_v < QOP

Figure 2-6. Solar position angles used to determine the incident angle.[13]

The general formula used to calculate θ is written as follows:

$$\text{Cos } \theta = \text{Cos}\beta \text{ Cos}\gamma \text{ Sin}\xi + \text{Sin}\beta \text{ Cos}\xi^{13}$$

You will remember β as the angle designated solar altitude. The angular difference between another familiar term, solar azimuth θ, and the azimuth of your mounting wall ("wall azimuth") ψ (psi) is needed to determine the γ (gamma) term in the equation. This term is imaginatively called the "wall-solar azimuth" (Figure 2-5) and simply quantifies the compass heading of your house with respect to the compass heading of the sun. More formally, "for vertical surfaces facing east of south, $\gamma = \theta - \psi$ in the morning, and $\gamma = \theta - \psi$ in the afternoon."[13] The reverse holds true for a wall facing west of south. If the wall is facing due south, $\theta = \psi$ and $\gamma = 0$. Once the γ value for your situation has been established, the tilt angle remains as the only missing piece of information.

The ξ (sigma) term in the equation denotes that tilt angle. The best ξ to use for your situation will depend on the latitude of your location and the application for which your system is being built. Some useful rules of thumb have evolved among people in the solar energy field. The optimum tilt for flat-plate collectors (the most widely used device for residential applications) being used in a year-round mode such as domestic water heating, is an angle equal to that of the local latitude. A seasonal application such as winter space heating dictates a steeper tilt angle, to catch the lower winter sun. The rule of thumb gnerally used here is to use a tilt angle equal to the local latitude plus $15°$.

To dispell any doubts about using the equation for the incidence angle, consider this example. A house is located near Denver, Colorado, and has its southern wall facing $12°$ east of true south. The roof on this side of the house has a slope of about $40°$ with respect to a horizontal plane. If the owner of the house were thinking of installing solar collectors flush with this roof, how would he go about estimating the angle of incidence that would occur at, say, 11:00 A.M. solar time, March 21. The solution is really quite simple. Denver is situated at a latitude very close to $40°N$, so the solar position table in Appendix I for this latitude can be used with reasonable accuracy. From the description of the house and the desired time, we know that:

$$\beta \text{ (solar altitude)} = 47.7°$$
$$\phi \text{ (solar azimuth)} = 22.6°$$
$$\psi \text{ (wall azimuth)} = 12°$$
$$\xi \text{ (tilt angle)} = 45°$$

Since the collection surface faces east of south:

$$\gamma \text{ (wall-solar azimuth)} = \phi - \psi \text{ in the morning}$$

Therefore,

$$\gamma = 22.6° - 12° = 10.6°$$

Substituting these known values into the equation for the incidence angle:

$$\text{Cos } \theta = \text{Cos}\beta \text{ Cos}\gamma \text{ Sin}\xi + \text{Sin}\beta \text{ Cos}\xi*$$
$$= (\text{Cos } 47.7°) (\text{Cos } 10.6°) (\text{Sin } 45°) + (\text{Sin } 47.7°) (\text{Cos } 45°)$$
$$= (0.673) (0.983) (0.707) + (0.740) (0.707)$$
$$= (0.468) + (0.523)$$

$$\text{Cos } \theta = 0.991$$

Therefore,

$$\theta = 7.62°$$

This answer means that the collectors would be less than 8 ° away from pointing directly at the sun at this time and location. This exercise only allows you to determine the incidence angle for specific conditions, and does not provide enough information to determine whether the collectors should be mounted flush with the roof for maximum performance.

The following rules of thumb for tilt angle would come in here, because they are based on many incidence angles averaged over the long term. For this house, if those collectors were to be used for domestic water heating, the optimum tilt angle would be equal to the local latitude 40°. If the collectors were to be used instead for space heating purposes, the tilt angle should be 15° greater, or about 55°.

Again, the θ value for the system should be kept as small as possible for a maximum return on your investment. This one small piece of information created the need to determine all those other angles. You simply want to know which way to position your collectors for the best performance. Discussions in this chapter were directed specifically toward knowing how much sunlight one can expect to fall on a system, and which way to point to catch most of it. Chapter 3 discusses that which you will be pointing.

---

*Do not let the appearance of this equation confuse you. As can be seen in the example problem, solving for θ requires only simple substitution and standard trigonometric relationships.

CHAPTER 3

THERMAL COLLECTION DEVICES

The reader should now have at least a rough idea of the intensity and character of incoming solar energy, the huge distance over which it has traveled, and its changing angle of approach. The idea of ever-changing energy availability has been discussed. This and subsequent chapters to it endeavor to show the major techniques and equipment now being developed or already in use to capture that energy and to convert it into forms more suitable to earthly applications.

For millenia, man has used energy deliberately released from organic fuels (fires of wood, coal, sod, manure, etc.) to warm both himself and his immediate environment (cave, hut, house). Ever since he first discovered that it was possible to alter his surroundings and make them more livable, he has been trying to do it more effectively. Until recent times, however, man was limited in his tailoring of his environment, with the fireplace, stove and chimney as his most sophisticated heat source and distribution equipment. The sun as a source of residential heat was thought of little or not at all by many cultures, and much of what was learned concerning functional architecture (window placement, house orientation, use of materials, natural ventilation, etc.) was later forgotten.

The knowledge was forgotten or deemed obsolete within the last century because of the effectiveness of recently developed central heating systems and the still more recent availability of an extremely convenient form of energy—electricity. With the new flexibility afforded by cheap, seemingly limitless supplies of electricity, oil, coal and natural gas, no need was seen to ever again have to design housing according to nature's dictates. The attitude that has evolved simply prescribed a larger furnace or air conditioner to remedy a too-hot or too-cool living space. In other words, technology is being used in brute force fashion to compensate for mediocre housing design. Economics did, and still does, reign supreme in these matters, with low energy prices making energy-wasting buildings appear cost-effective. This short-sighted attitude is proving to be short-lived, however. The energy sources just mentioned are anything but limitless, and their formerly low costs are rapidly and permanently becoming a thing of the past. The human species now is confronted by a new set of survival alternatives. We can continue to think of nature as an adversary to which compromises must now be conceded; or we can finally realize that the natural environment is something we must learn to understand, and that it offers benefits we can use as well as the harshness we can avoid.

A wealth of ideas concerning this necessity for a new energy ethic were put forth during the World Game Workshop held in Pennsylvania during the summer of 1974. Perhaps the notion paramount in importance was the classification of energy sources into two main groups: "capital" energy sources and "income" energy sources. The concept, like most important concepts, is a simple one. "Capital energy sources are finite in quantity and cannot be replaced once used. Income energy sources are regenerative and are limited by rate of use rather than supply."[3] When you consume a barrel of oil, a ton of coal, 1000 ft^3 of natural gas or an ounce of uranium (all "capital" forms of energy), you cannot, obviously, call them back into existence. They have ceased to be. To satisfy your continuing need for energy, you must go back into the field and dig up more. After a sufficient length of time, the resource either becomes depleted or, more realistically, becomes so expensive to locate and extract that it may as well not exist. This will inevitably happen with all capital sources of energy. The only variable is time.

E. F. Schumacher has been quoted as saying that "nothing makes sense unless its continuance for a long time can be projected without running into absurdities. . . ."[16] Although generalities are impossible to prove, the quotation succinctly states an intuitively sound principle. To seriously base a long-term energy philosophy on resources that will run out in the historically near future could easily be construed as an absurdity.

An income energy source is a wholly different kind of entity, however. Instead of dipping into a fixed reserve of materials, income energy is extracted from natural ongoing processes. As stated in Chapter 2, most of these processes are driven by solar radiation. Among the various income energy sources available to us are those related to the geosphere (solar power, hydroelectric, wind power, wave power, ocean thermal gradient, geothermal, etc.) as well as the biosphere (algae/bacteria methane production and wood farming). Man would be hardpressed to use up even a significant fraction of this huge amount of energy. Even if he did, regeneration of that energy would be immediate and automatic.

This concept can be applied to our original discussion of man's efforts to tailor his immediate surroundings to better suit his body's requirements. A very large portion of the processes by which man utilizes energy are based upon energy in the form of heat. In most applications, some working fluid such as air, water, a refrigerant or some chemical compound is heated to some operating temperature and distributed to some location in a system at which it can perform work. The heating process has traditionally been accomplished by combusting some fossil fuel to directly or indirectly (as in the case of centralized generation of electricity) heat the working fluid. This specifically applies to residential uses such as space heating and domestic water heating, as well as the innumerable processes used by industry.

It has been estimated that almost two-thirds of all residential and commercial energy consumption in this country is expended toward heating occupied space and to produce domestic hot water.[17] This is, of course, representative of a large expenditure of energy. A fortunate characteristic of both of these applications, however, is that both objectives can be satisfied by using heat at moderately low temperatures (below 200°F). It is precisely this fact that makes space and domestic water heating the two most cost-effective solar energy applications at the present time.

Aside from its variability, the prime technical difficulty in harnessing sunlight is its diffuse nature. It does not come in concentrated easily stored forms as do the fossil fuels. However, at average levels of solar flux in this country (and in all temperate regions of our planet), the temperatures necessary for space heating and domestic water heating can be achieved a significant portion of the time with sunlight captured by simple, nonconcentrating, stationary equipment.

NONCONCENTRATING COLLECTORS–FLAT-PLATE TYPE

Every material in existence is composed of atoms and molecules that vibrate in characteristic patterns, with characteristic intensities. These patterns and intensities are what enable us to tell different materials apart. Adding energy to a material causes these molecules to become more active and vibrate with greater vigor. The extent to which the molecules of a substance are in motion exactly corresponds to how much heat is present within the substance at that instant. As more energy is pumped into a unit mass the "heat intensity" increases, raising the temperature of that mass. Thus, it can be seen that temperature is a measure of heat intensity, and that adding energy to a substance increases its temperature.

By the time sunlight gets through our atmosphere, as discussed in Chapter 2, it consists mainly of visible light and infrared radiation. When the sunlight strikes an object, a portion of the energy bounces off (reflection). The remainder of the energy is absorbed by the object, and serves to increase its molecular activity, causing a corresponding rise in temperature. For a given intensity of insolation, a material will heat up a certain amount, depending on its color, surface texture, shape, conductivity, orientation and other factors.

The preceding discussion shows the fundamental simplicity of the sunlight-to-thermal energy conversion process. The conversion takes place automatically on contact between sunlight and a material. The trick is to maximize the efficiency of the process, and to transfer the resultant heat energy to where it is needed. To be able to utilize the sunlight-to-thermal phenomenon effectively requires a device capable of heating up efficiently, able to transfer the collected heat to some kind of heat transfer fluid (either air or a liquid) and be able to do this while exposed to outdoor weather conditions. The most popular device at present capable of fulfilling these requirements, is called a flat-plate collector. While tremendously improved, today's collectors are direct descendants of the prototypes used by Willsie and Boyle and those used in the M.I.T. Solar House I during the first half of this century (see Chapter 1).

There are dozens of flat-plate collectors of different design commercially available at the present time. The rapidly growing demand for collectors, the vast size of the potential market and the variety of specific uses to which the devices may be put have spurred numerous design approaches to be taken by various researchers and manufacturers. New ideas are being investigated daily. In spite of this, the vast majority of flat-plate collectors have five basic components in common (Figure 3-1):

1. the absorber surface;
2. the heat transfer interface/fluid passage;
3. the glazing;
4. the insulation; and
5. the protective casing.

LIQUID-COOLED FLAT-PLATE COLLECTORS

Many of the principles involved in the operation of the liquid-cooled collectors are applicable to the air-cooled collectors as well. It will be helpful to keep this in mind as the discussion develops.

The first component listed above is the absorber surface. This is the part of the collector that actually collects the sunlight and converts it to heat. While the surface itself may have corrugations, grooves, dimples or seams, it is still basically a flat plate in appearance, and does no concentrating of the incident sunlight (as do curved surfaces). This is the

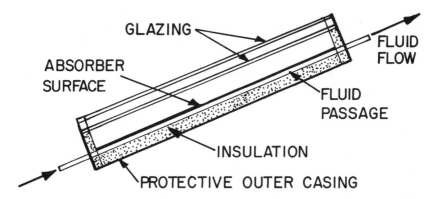

Figure 3-1. The five basic components of all flat-plate collector panels. While the shape and arrangement of these units may vary, all functions must be fulfilled for proper operation of the device.

characteristic that gives the device its name. Most collectors utilize plates made of metal (copper, aluminum or galvanized steel), but some are fabricated of plastic for lower temperature applications such as swimming pool heating. The top surface of the absorber plate will have either a flat black coating or a selective surface applied to it (to be discussed further on), to maximize the amount of sunlight absorbed and minimize the amount of the heat emitted by the plate. Were the absorber plate to be exposed to sunlight by itself, it would simply heat up to some equilibrium temperature at which the energy gained from sunlight were equal to the energy lost to the surrounding air through radiation and convection.

Thus, the absorber plate alone is of little use, and must be cooled for two reasons. First, the heat energy collected by the device must be transported to where it can be used (inside the dwelling). Second, the efficiency of heat collection drops drastically at high temperatures. This latter point can be thought of as being reasonably analogous to a sponge absorbing water. The sponge can absorb only so much. It can easily be seen that the sponge must be squeezed periodically, so that it may soak up another batch of water. In like fashion, heat must constantly be removed from the absorber plate if the plate is to be able to continuously absorb more.

In the liquid-cooled collector, water or a water/antifreeze solution are used predominantly as the heat transfer (heat removing) fluid. Nontoxic antifreezes and silicon-based heat transfer fluids are also being developed specifically for use in solar energy collection systems.

There are a number of ways in which the heat transfer fluid can be brought into contact with the collected heat. The simplest of these is the "trickle-type" arrangement first used by Dr. Harry Thomason for a house he built in Maryland in 1959. His Solaris collectors use a blackened corrugated metal sheet as the absorber surface, which is oriented toward the south at some tilt angle. Figure 3-2 illustrates the appearance of this device. As can be seen, the heat transfer fluid flows through holes in a header pipe located along the top of the corrugated sheet, trickles

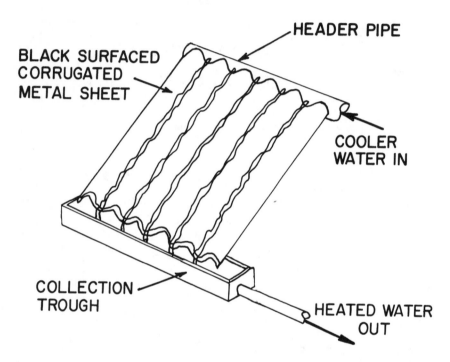

Figure 3-2. The trickle-type absorber designed by Dr. Thomason. This is a schematic representation, in that the unit is also glazed, insulated and surrounded with an outer casing. As can be seen, this type of collector cannot be incorporated into pressurized piping systems.

down the "valleys" in the sheet and absorbs the absorber's heat as it goes. A collection trough at the bottom of the collector routes the flow back to a return pipe. The design incorporates simplicity with good performance for outlet temperatures of 110°F or less, "but the collection efficiency is caused by poor heat conduction from the ridges of the corrugations to the valleys, and an additional heat loss is caused by the convection of saturated air above the absorber plate.[18] In this type of collector, the heat transfer surface is the area of the plate wetted by the transfer fluid. External to the collector, the fluid is distributed through conventional piping, a feature common to all liquid-cooled collection systems.

The next type of absorber is a logical evolution of the trickle-type and can be characterized as the "tube and sheet" configuration. Parallel metal tubes of fairly small diameter (3/16-1/2 in. o.d.) are clamped or soldered to the top or bottom surface of the absorber sheet, at spacings dictated by the designer (4- to 8-in. on centers). As shown in Figure 3-3, the transfer liquid usually is fed into the lower header pipe, and the flow is thus split up to feed each individual tube. The flow proceeds upwards through each tube, heating up as it goes. The header at the top of the collector serves to gather and mix the individual flows and to direct the fluid back into the distribution piping circuit. Good performance is shown by this type of collector panel. Heat losses due to convection are minimal, and the absorber plate coating is not subjected to the erosion-corrosion effects of the flowing liquid as is the coating in the trickle-type collector. The major disadvantage inherent in the tube and sheet arrangement is encountered during fabrication, since soldering, clamping or brazing each tube to the absorber plate is crucial, time-consuming

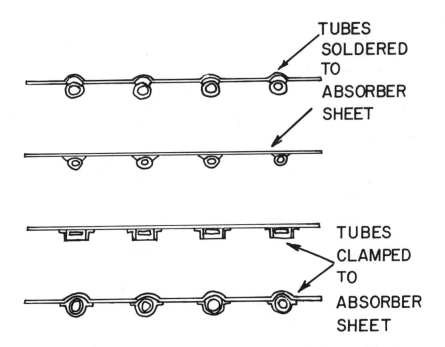

TUBES SOLDERED TO ABSORBER SHEET

TUBES CLAMPED TO ABSORBER SHEET

Figure 3-3. A few examples of tube and sheet absorber plates, in which the tubes and absorber are formed separately and assembled into a composite structure. (Note: These configurations all have the tubes below the sheet and could, of course, be inverted).

and costly. This type of configuration is difficult to reconcile to mass production techniques, although some manufacturers have minimized many of the problems. This type of collector panel is, however, definitely the most successful of the first generation of commercial liquid-cooled, flat-plate collectors.

A variation on the tube and sheet arrangement is the Finplank assembly manufactured by the Sunstream division of Grumman, which consists of an absorber plate made of interlocking aluminum strips. At the edges of each strip is a flange designed to mate with the adjoining edge of the adjacent strip. Incorporated into this linear mechanical joint is provision for a small diameter copper tube, as shown in Figure 3-4. Thus, when assembled, this absorber plate has flow tubes locked into its own structure, yielding an assembly that may be deemed a compromise between the tube and sheet configuration and that of the "integral fluid passage" type of absorber. Performance is quite similar to the tube and sheet-type collector, but the step of soldering or brazing the tubes is eliminated.

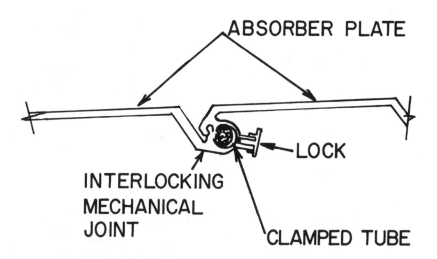

Figure 3-4. The Finplank absorber, composed of a series of strips that interlock at joints like the one shown above. The absorber plate is aluminum, the tubes are copper.

The third major category of heat transfer arrangements is the integral fluid passage-type of absorber (Figure 3-5). In this case, the absorber plate itself has built-in passages for the transfer fluid. Two commercial versions of this configuration are Roll-Bond panels manufactured by the

Olin Brass Company and the Tube-in Strip Collector by Revere Copper and Brass. This type of absorber plate lends itself to lower mass production costs and provides excellent heat transfer performance when in operation. Once the fabrication technology for making this type of collector plate is developed sufficiently, this configuration will very probably become the most widely used of the flat-plate collectors.

Figure 3-5. The integral fluid passage type of absorber plate, in which the tubes and the plate are formed simultaneously. The cross-sectional shape of the tubes varies from manufacturer to manufacturer.

A "sandwich-type" absorber was developed during operation of the University of Minnesota's Ouroboros House by John Ilse, one of the graduate students involved in the project. The original plans for the house were generated by Dr. Thomason and subsequently "modified by the student builders for Minnesota's extreme cold."[18] Originally, Thomason's collectors were used, but Ilse's sandwich panels later replaced half of these. "Projected total output values . . . indicate that Ilse's internal-flow steel-sandwich collector was operating between two and three times more effectively than the open-flow trickle type. . . ."[18] In constructing the sandwich-type absorber, two metal sheets are dimpled in corresponding locations and spot-welded together. The sheets are then welded along their perimeter. The resultant panel (Figure 3-6) is fitted with a supply pipe at one end and a return pipe at the other. Heat transfer liquid is then able to flow through the sandwich, soaking up heat abosrbed by practically the entire surface area of the collector. Excellent heat transfer behavior is exhibited by this arrangement, and low square foot fabrication costs are expected if and when mass production is established.

It should be emphasized that we have only discussed the main absorber plate/fluid passage designs. Many other variants exist, but most fall either into one of the previous categories or somewhere between them. The probability is near zero that any of the designs currently available will not be modified again and again as the state-of-the-art advances. The important thing to retain here is a feeling for what the absorber plate and its fluid passages are supposed to accomplish, so that exposure to new designs or closer examination of existing ones will be interesting rather than mystifying. Reiterating then, the functions of the absorber plate and fluid flow passage components are these:

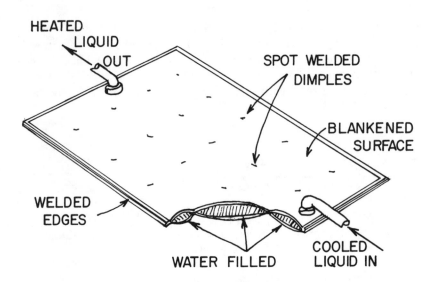

Figure 3-6. A rough sketch of the type of absorber designed by John Ilse—the sandwich type. Although this type can be made in any good welding shop, care must be taken to ensure water tightness.

1. to absorb a maximum amount of sunshine and convert it into thermal energy with a maximum of efficiency; and
2. to transfer as much of the collected heat to the chosen heat transfer fluid as possible, so that the heat may be transported to point of use, and to allow the now cooler plate to absorb more incoming sunshine.

In addition to these theoretical goals, there are "real-world" goals that are desirable as well. Designers of these components are also trying to achieve minimums of: hydraulic pressure loss through the collector, the likelihood of leaks or mechanical failure, cost of material, cost of fabrication, weight of collector, differences in fluid flow between any two passages (which would lead to "hot spots") and any difficulties involved with assembly, disassembly or repair of the collector.

To increase the efficiency of these absorbers, their surfaces are coated with various substances capable of absorbing energy. Just as a black object gets hotter than a white object when both are exposed to the sun, so too are some surfaces better able to absorb energy than others. The percentage of sunlight absorbed by a surface is known as its "absorptance" (α). The other side of the coin, the amount of heat that the warmed surface radiates or gives off, is quantified by the term "emittance" (ξ). The higher the emittance the more heat is lost from the surface. The maximum value of ξ is 1.

For the purpose of collecting solar energy, the best kind of surface would be one that would absorb a very large fraction of the energy incident on it (yielding a high α value) and also emit a very small portion of the heat it has collected (yielding a low ξ value). Flat black paints are very helpful in attaining high absorptance, but also have high emittance values. This becomes important only at higher operating temperatures, so inexpensive black paints can be used very successfully as collector coatings when the system is to operate at 100°F or below.[12]

Combination of high absorptance and low emittance can be obtained, however, and are provided by a group of substances called "selective surfaces." Since longwave radiation (with wavelengths greater than 3.0 μ) comprises only a small fraction of incoming solar energy, and the heat energy that would be radiated by a heated surface would be longwave radiation, an interesting possibility arises. A material that reflects very little of the incoming solar energy with wavelengths less than 3.0 μ and emits very little longwave radiation would be an ideal surface for solar energy collection.[14] Since such a surface would be tailored to selectively absorb certain wavelengths and emit others, the same selective surface is an appropriate one. These surfaces allow the collectors upon which they are applied to operate at higher efficiencies and at higher temperatures than if flat black surfaces were used.

There is quite literally, however, a price to pay for these benefits. They are very costly, with additional problems being caused by the need for extremely accurate quality control during manufacture and problems with sustained durability.[19] These surfaces are many, composed of different substances, and are applied in different ways to different metals. They are also experimental, and very few are supplied on commercially available absorbers without a substantial increase in cost. As time goes on, the performance and cost-effectiveness of these surfaces will improve, but the requirements of a system may or may not immediately justify the increased cost. They are not, as yet, the automatic option to choose.

Let's assume, however, that the optimum types of absorber, absorber surface and fluid passage have been chosen. Another problem that the designer must face is how best to allow the sunshine to reach the absorber plate while preventing rain, snow, wind and cold outside temperatures from harming the plate surface and stealing away collected heat. The obvious answer is the correct one: cover the absorber plate with a transparent material capable of transmitting sunlight, yet tough enough to withstand prolonged exposure to the elements. This component is known as the glazing and the materials used are glass, plastics, fiberglass or combinations thereof.

Glass for this application fulfills the requirements of transmission and toughness while providing an additional benefit. Glass is "almost totally opaque to thermal (longwave) radiation."[12] At first, this may sound like a disadvantage, but most of the longwave radiation trying to pass through the glass would be formerly shortwave radiation trying to escape from the absorber plate after its conversion to heat. Thus, the glass, in addition to preventing moisture from entering the collector and air convection heat loss, would also serve to form a longwave heat trap, which helps the collector to operate at a high efficiency. This heat trap phenomenon is more generally known as the "greenhouse effect," and explains the reason for the greenhouse's success.

Glass is also very resistant to scratching and practically impervious to the damaging effects of ultraviolet exposure. High temperatures (maximum stagnation temperatures in flat-plate collectors seldom exceed 400°F) do not adversely affect this material, another plus in its favor. "Depending on the iron content of the glass, between 85 and 92% of the sunlight striking the surface of a 1/8-in. sheet of glass (at vertical incidence) is transmitted."[12]

A list follows of disadvantages of glass. While a low iron content is helpful to the transmittance of glass, low iron glass is more expensive. Another factor to consider here is the brittle nature of glass, if not properly tempered. The tempering process, while adding much to the necessary toughness of the cover glass, adds again to its price. In fact, the cost of a good quality, fully tempered sheet of glass suitable for use in a solar collector may run as high as $2.50/ft^2, or more. Add to this the high weight of glass relative to the weight of plastic, and it becomes evident that glass has problems as well as advantages.

The plastics are not perfect either. They are lighter, stronger and usually cheaper than glass. They can equal or exceed glass in its transmittance capability, and can be fabricated in larger sheets. However, most plastics are not opaque to longwave radiation, eliminating the heat trap phenomenon's boost to high-temperature efficiency. Most plastics also have a decided vulnerability to the scratching effects of dirt, dust or sand. Ultraviolet radiation yellows and weakens many plastic materials after one or two heating seasons, making their replacement a significant part of the maintenance costs involved in a solar collection system using them. Attempts are being made to develop a plastic or fiberglass material more suited to the harsh life of a collector cover, and the near future may see some such material become easily available. Until that time, however, plastics simply do not measure up to high-quality glass in the reliability, optical, thermal and dimensional stability, and scratch resistance required for solar space heating and domestic water heating systems. One is probably better off with good glass.

Choosing a material is not the only decision involved in correctly glazing a collector. The optimum number of cover sheets to use must also be determined. The temperature at which the collector will operate is the prime factor here, with outside weather conditions being second in importance. For example, if the collector will be used only to heat swimming pool water, a relatively low operating temperature will be required, in conjunction with warm ambient weather conditions. Thus, it would be necessary to use only one cover plate for the collector, and it is possible that complete elimination of the cover plate could be acceptable.

Space heating and domestic water heating systems operate at higher temperatures, however, and outside air temperatures during the heating season can get quite low. Under these conditions, the heat loss from the collector can be very large. A mathematical model was developed by Löf and Tybout, to predict the thermal performance of flat-plate collectors in which, logically, many parameters were involved. A summary of their findings stated "the best number of (ordinary) glass covers was found to be two for all locations except those in the warmest and least severe climates...."[14] When "best number" is used, it refers to the number of covers that yield the least expensive energy from the solar collection system. In most areas of the country, any money saved on using less glazing would be more than offset by the additional expenses incurred for the larger equipment you would need to store and distribute a given amount of heat at the resultant lower operating temperature.

The final components of the flat-plate collector are the insulation and protective outer casing. Just as the cover plates reduce heat loss out of the front and raise the collector's efficiency, insulation is installed behind the absorber plate and along its edges to further contain heat. When the collectors are installed flush with the roof of a building, any winter heat loss through the back or edges of the collector are directed into the building, a fact which would seem to suggest that insulation is not necessary in this case. There are factors not immediately evident, however, that override this argument. The first is that the heat lost to the building stays near the roof by natural convection, doing little towards heating occupied space. Secondly, lower output temperatures would cause the same equipment problems associated with insufficient cover plates. If the system is in operation during the summer months, as in a domestic water heating system, even more heat would be directed into the building than during the winter, which is hardly desirable. As with the house itself, then, insulation is both necessary and desirable, to keep the heat energy in its proper place.

Finally, the protective outer casing literally serves to hold the entire device together, while supplying whatever weatherproofing and structural integrity are necessary for the finished product. Most commercial collectors use galvanized steel or aluminum for their casings, and the surface finishes desired will vary according to what corrosion resistance is necessary for a particular area.

The device is now complete in and of itself. It need only be mounted at the correct orientation, plumbed, filled with fluid and exposed to the sun to start earning its existence. As a rough rule of thumb (specific statements apply only to specific systems), most liquid-cooled collectors are set up to operate at temperatures of 75-100° F above ambient outside temperatures. Efficiencies of about 50-60% can be achieved under these conditions, with greater efficiencies at lower collector temperatures. One should look for efficiencies of at least 45-50% for a system. The uninstall-ed costs of the collectors, with double glazing and selective surface, should run around $15/ft^2 or less. Again, prices will also vary widely, but should improve with time. As is obvious, the design objectives of a system will dictate what trade-off decisions must be made. Most decisions have already been made by the manufacturer, and common sense and basic understanding will dictate the rest.

AIR-COOLED FLAT-PLATE COLLECTORS

The relative merits of liquid-cooled and air-cooled flat-plate collectors are defined by both physics and economics. For the same rise in tempera-ture, a unit volume of water can absorb more than four times as much heat than can air. This makes it necessary to supply air to air-cooled collectors with much larger fluid passageways (air ducts) than would be necessary to supply liquid to liquid-cooled collectors (piping), since much larger volumes of air must be circulated. This raises the cost for air sys-tems and may cause problems with available space.

While the "liquid-cooled collector can be used for space heating, domes-tic water heating, pool heating and space cooling..."[20] the air-cooled col-lector is not quite so versatile. This inequity is explained by the fact that the heat exchange surfaces necessary to transfer a unit amount of heat between air and liquid are much larger than the heat exchange surfaces necessary to transfer the same quantity of heat between two liquids. The differences in the heat capacities and fluid mechanics of the two substances are responsible for this situation, and make air-to-liquid inter-faces larger and more expensive than the liquid-to-liquid type.

On the other hand, the same physical constraints that cut down the versatility of the air-cooled collector give it a decided advantage for space heating applications. Since liquid-cooled, flat-plate collector output tem-peratures during the heating season would be somewhere between 100 and

130°F, depending on insolation levels and system operating conditions, a problem arises if the liquid's heat is used to supply a baseboard heating system. Normal baseboard heaters require fluid temperatures of approximately 180°F to work effectively. To supply the same amount of heat to the living space with fluid temperatures of about 120°F, the amount of baseboard length necessary would be increased greatly. In addition, the liquid that flows through the collector system should be circulated through a heat exchanger, giving its heat to the water flowing through the baseboard (hydronic) heating system. The losses always encountered in heat exchangers would further decrease the fluid temperature available for space heating purposes.

The output temperatures available from air-cooled, flat-plate collectors would be very similar to those available from the liquid-cooled type, but with an important difference. If a forced air system is used to heat the space within a building, the heated air is blown directly into the rooms, instead of circulating through tubes along the walls, as in the hydronic system. Therefore, the output air from the collectors can be used directly to heat the building. Any increase in the size of the air ducting over a "conventional" system would be minimal, and no intermediate heat exchange would be necessary.

These, and other technical and economic factors (discussed in Chapter 4) can determine whether to use liquid or air collectors. To be better able to judge, an understanding of air-cooled collectors would be helpful. As is the case with the liquid type, air-cooled collectors must incorporate five basic components, including the absorber surface, heat transfer interface/ fluid passage, glazing, insulation and protective casing.

With air-cooled collectors, many fabrication steps are not as crucial as with liquid-cooled units. Working pressures are much lower, leakage is of minor importance, freezing problems are eliminated, and the heat transfer liquid in air-cooled collectors can be directly brought into contact with the entire heat absorption surface. This last factor is of major importance in air collector design. Since the fluid need not be restricted to pressure-tight tubes, with heat transfer totally dependent on conduction through metal barriers, many different absorber surface configurations are possible. Since air can extract heat from any sun-heated surface with which it has turbulent contact, the surface itself does not have to rely on metal's high thermal conductivity to be effective and can, therefore, be made of other materials. Metal is, however, still useful in eliminating any "hot spots" caused by uneven air flow. Although air-cooled collectors have received far less engineering attention than the liquid type, many configurations have been fabricated and tested. Black stretched cloth, corrugated metal plates, finned metal plates, perforated absorber plates, rough-surfaced flat plates and even black-painted glass sections have been used in these investigations (Figure 3.7).

The air has been channeled to flow above, below, or on both sides of, flat plates (as a simple study case) yielding valuable information on the relative performance of each arrangement. In general, at space heating temperatures, the highest efficiency is obtained when the air is circulated below the absorber plate, leaving a layer of still air between the absorber plates and the cover glass.[21] Turbulence is another important factor in the efficient operation of these devices. If the surface of the absorber plate is too smooth, or the air flow is too slow, a very thin layer of still air "coats" the absorber surface, serving as an obstacle to heat transfer. To overcome this effect, the absorber plate should have a rough surface, with many irregularities and sharp edges. Air flow should have a sufficient velocity across that irregular surface to create the necessary mixing action to break up the still air layer. As with everything else, however, this can be carried too far. If the resistance to air flow is increased excessively, a larger fan may be needed to circulate air through the system. The energy requirements of this fan may outweigh any increased solar efficiency. General ranges of air flow and air velocity will be discussed in Chapter 4.

It has been shown that the absorber plate itself also serves as the heat transfer surface and partly as the fluid passage. In addition to the ability of a metal absorber to reduce "hot spots" in the collector, it is also the only material that can be used if a selective surface is desired. Thus, metal (copper, aluminum or galvanized steel) is still preferable to the other materials. This may change as technology advances.

The rest of the fluid passage can be provided in a number of ways, but the least expensive would be a foil layer applied to the inside surface of the edge and back insulation, providing both an airtight fluid path and an added barrier to thermal losses.

Insulation, glazing and casing requirements are not quite as stringent, from the point of view of fluid leakage, as they are with the liquid-type collectors. Quality should be kept as high, however, for purposes of efficiency and durability. All in all, there are no major differences in the costs of the two types of collector. Internal duct sizing experience usually dictates air distribution ducts of approximately 0.5-1 in. in depth for collectors using flat absorber plates,[12] with larger ducts necessary for collectors that use natural convection instead of forced flow. The optimum size of these ducts varies, depending on the collector design and application, and represents a workable compromise between maximal heat transfer (turbulence) and the resultant increase in pressure drop through the collector. While a number of firms are presently manufacturing air-cooled, flat-plate collectors, the selection is not as wide as that available with

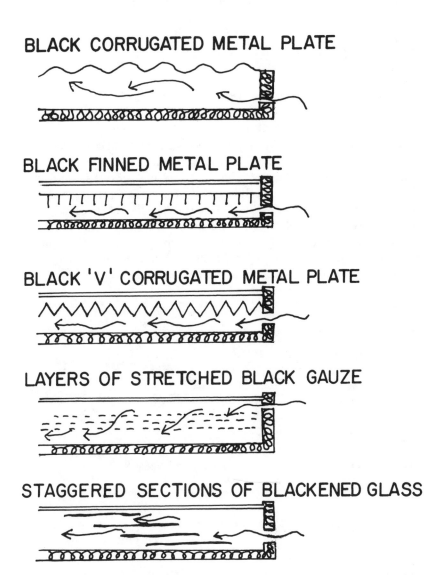

Figure 3.7. These are but a few of the many possible arrangements of various materials that have been and are being used to form the absorber of an air-cooled, flat-plate collector.

the liquid-cooled units. As more attention is paid to the potential uses of air-cooled collectors for space heating, this selection will expand.

NONCONCENTRATING COLLECTORS –
EVACUATED TUBE TYPE

While flat-plate collectors are used almost exclusively at present to actively collect solar energy for low- and medium-temperature applications, a new type of nonconcentrating collector is breaking into the business. It is cylindrical in shape, rather than rectangular, but basically uses the same optical and thermal principles to do its job. The device is called an evacuated-tube collector. It, too, can use liquid or air as its heat transfer fluid, and is shown in cross section in Figure 3.8. This particular device is being developed by Owens-Illinois and is called the Sunpak Solar Collector. The description that follows applies specifically to this item, although similar devices being produced by Corning Glass Works, Argonne National Laboratories, General Electric and a European firm, N. V. Philips Gloeilampenfabrieken,[22] operate in generally the same fashion.

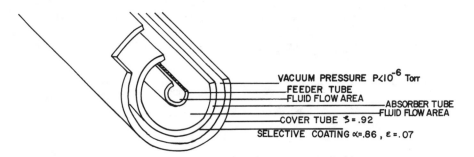

Figure 3.8. Shown here in cross section, the Sunpak evacuated-tube collector is able to achieve higher operating temperatures than can flat-plate collectors.

As seen in the cross section, the collectors incorporate two concentric metal tubes situated within a glass tube. The small central tube acts as the feeder tube, and brings the heat transfer fluid into the collector. The fluid undergoes some preheating before it reaches the opposite end of the tube. At the opposite end, the flow reverses direction and flows back towards the inlet end, in the space between the feeder tube and the absorber tube. It is along this pass that the majority of heat transfer takes place. The fluid is then discharged to a manifold, which routes the flow to the next tube, where the process is repeated. In this way, each tube serves to increase incrementally the temperature of the fluid.

The outer glass tube is 2 in. in diameter, about 40 in. long, and serves as outer casing, glazing and insulation for the collector. The casing and glazing functions are fulfilled because the tubes "are made of low-iron borosilicate glass, chosen because of its ability to withstand weathering as well as stress associated with high temperatures."[23] A vacuum of approximately 10^{-4} Torr (less than one-millionth of atmospheric pressure) is drawn between the glass tube and the absorber tube. This is the factor that provides the insulation capability, since the vacuum is an effective barrier against conductive and convective heat loss.

The outer surface of the absorber tube is provided with a selective surface (α=.86, ϵ=.07),[24] to boost performance efficiency at high operating temperatures. In addition to the vacuum, the shape of the device is a major point of difference with respect to flat-plate collectors and allows sunlight to enter the tube from essentially all directions. The combination of selective surface with vacuum and shape allow the Sunpak units to achieve the 190-240°F output temperatures necessary for driving absorption type cooling equipment.

A total of 24 tubes are assembled into each module, with a central connecting manifold, and the assembly measures approximately 4 x 8 ft (Figure 3.9). Each module has an effective collection area of some 27.4 ft^2 and weighs about 185 lb when filled with water.[22] As installed, the tubes are parallel to each other, spaced about 2 in. apart, and supported about 6 in. from the reflecting surface on which the collectors are mounted (Figure 3.10). This arrangement is used to take advantage of the 360° aperture of the tubes. The light passing between the tubes bounces off the reflecting surface in diffuse form and is directed back to the tubular absorbers. According to the manufacturer, the tubes "collect the diffuse component of insolation efficiently due to their curved surfaces, which intercept more diffuse light from the sky dome than a flat surface with the same projected area."[24] A curved, lightweight aluminum reflector, able to boost the energy output of the collector significantly, is also available.

The cost of these units is approximately $548 per module before installation, with the added cost of the aluminum reflector increasing this figure by about $68 per module.[22] The units are currently being used on solar projects at two facilities, including the Terraset Elementary School in Reston, VA (4,700 ft^2) and the Federal Office Building in Saginaw, Michigan (7,000 ft^2).[24] The collectors provide heating and air conditioning energy at both installations. More operating data is needed for these devices to gain widespread acceptance, but they appear to hold much promise.

Figure 3.9. The end of the evacuated tube is sealed such that the vacuum between the glass and absorber tubes is maintained for the life of the collector. In the background, installed 24-tube modules can be seen with their manifolding connected in a series configuration. (Photo courtesy of Owens-Illinois.)

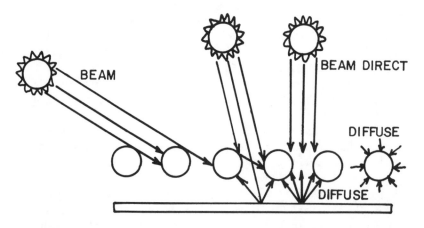

Figure 3.10. The $360°$ aperture of the evacuated tube collector allows radiation to strike the absorber tube from essentially all directions.

CONCENTRATING COLLECTORS

For applications that require temperatures in excess of those achievable with the devices discussed thus far, sunlight must be concentrated to an intensity greater than that naturally available. In essence, a concentrating collector employs either refraction (lenses) or reflection (mirrored or reflective surfaces) to channel natural concentrations of energy falling on an area into a significantly smaller area, the focus of the system. In this way, the energy intensity at the focus is much greater than that of natural sunlight, allowing the absorption surface or solar cell reception area (see Chapter 4) to operate at much higher temperatures. The area that receives this augmented insolation is called the "hot spot" and may be a small circle, a rectangle, a thin line or some other shape, depending on the techniques used.

The extent to which concentration is accomplished is called the concentration ratio (C) and is generally expressed:

$$C = \frac{q_f}{q_i} = \frac{\text{augmented insolation at hot spot}}{\text{unconcentrated direct insolation}}$$

For example, a device having a C of 4 would have a hot spot that is receiving solar energy that is four times as intense as that which a flat-plate collector would receive, without concentration, at the same location. The optical price paid for these increased intensities is a system that depends more on pointing directly at the sun. As the concentration ratio increases, the "acceptance angle" (amount of deviation from perpendicularity acceptable) decreases, and it becomes more and more necessary to allow the device to follow the sun's movement across the sky. In other words, devices that provide low C-factors can operate with minimal or no adjustment of orientation, while high C-factor devices require increasingly more elaborate and precise sun-tracking capability. Diffuse sunlight can be collected with low-concentration devices, but its contribution to high-powered concentrators shrinks to insignificance very quickly, making devices with high C-factor devices.

The simplest and least expensive form of solar thermal concentration is achieved by placing flat (planar) reflective surfaces (sheet metal, foiled or white surfaces, etc.) near an array of flat–plate collectors, as shown in Figure 3.11. This technique has been used in numerous installations and can as much as double the amount of sunlight striking the collector. Diffuse as well as direct-beam radiation can be concentrated in this way. However, since both the reflector and the collector panel are mounted in a fixed position, this concentrating effect is significant only at certain

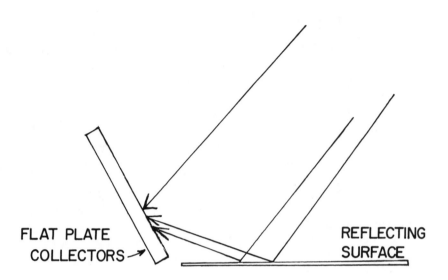

FLAT PLATE
COLLECTORS

REFLECTING
SURFACE

Figure 3.11. Some concentration of sunlight can be achieved simply by using a flat reflecting surface to direct sunlight from an adjacent area to the collection surface. The added input may or may not be worth the added expense.

times, and, thus, does not double the total energy collected by the system. The additional energy that plane reflectors do contribute is not consistently large, and so may or may not justify the added expense of building and installing them. The appropriateness of using them depends on location, materials used and system operation.

A different type of stationary concentrator is a device that employs a reflective sheet that is curved in two dimensions and focuses solar radiation into a line-shaped hot spot. Developed by Dan Lightfoot of the Alternate Energy Resources Company (AERCO) of Ottumwa, Iowa, the Hot-Line collector generally works as shown in Figure 3.12. Air, usually employed as the heat transfer fluid in this device, is forced through the wedge-shaped absorber channel, which is coated with black paint. Glazing, insulation and casing configuration are similar to flat-plate collectors. Although stationary, the reflective surface compensates for solar movement and can continually focus diffuse and direct sunlight onto the surface of the absorber pipe with good efficiency percentages (high 80s or low 90s).[25] The acceptance angle of the device allows it to successfully operate through a 45° vertical and a 150° horizontal sun displacement. While operating data are scarce, it appears that this device is one well worth researching further.

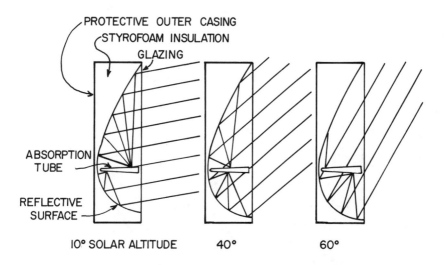

PROTECTIVE OUTER CASING
STYROFOAM INSULATION
GLAZING

ABSORPTION
TUBE

REFLECTIVE
SURFACE

10° SOLAR ALTITUDE   40°   60°

Figure 3.12. This simple schematic of the Hot-Line concentrator shows how its curved reflector serves to focus incident sunlight onto the wedge-shaped absorption surface. The shape of the components automatically compensates for a large amount of solar movement. The unit can also be mounted horizontally or at a tilt.

Curved, two-dimensionally reflecting surfaces have been used before, usually to focus energy onto a round tubular absorber pipe. This approach was used in the irrigation project designed by Shuman and Boys in 1912 (see Chapter 1). Many such configurations are in use today, but sun-tracking is necessary for optimum operation. Although this arrangement works well in direct sunshine, in conjunction with tracking about a north-south axis, the concentration ratio is usually high enough to render diffuse sunlight unusable. This factor makes a large difference in year-round totals of collected energy. For certain applications, however, the added problems and costs are offset by the higher temperatures and efficiencies reached during operation.

Another trough-like collector is being developed by the Northrup plant in Hutchins, Texas, but this one uses refraction to concentrate the sun-light. As shown in Figure 3.13, a curved Fresnel lens bends the light rays toward a focus located at the bottom of the "trough." A blackened metal tube situated at this focus is then able to transfer the received thermal energy to the liquid flowing through it. The casing of the unit is shaped as shown, with 1-in. thick fiberglass insulation situated along the inner surface. This sytem also employs north-south axis tracking (this means that the collectors can rotate to point from east to west during the course of the day), which is actuated by a small solar cell sensor connected to

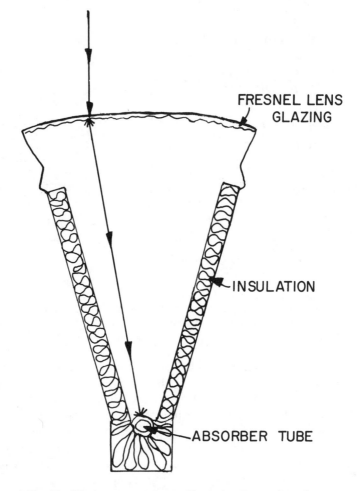

FRESNEL LENS
GLAZING

INSULATION

ABSORBER TUBE

Figure 3.13. The Northrup concentrator. Shown here as a schematic cross section, the manner in which the light ray is bent towards the absorption surface can be observed. This kind of concentrator must track the sun for successful operation.

an electronically controlled motor. Diffuse sunlight cannot be collected, but clear weather insolation is sufficient to allow the unit to supply energy to domestic water heating, space heating and even for the efficient operation of absorption-type cooling systems. Northrup prices these units at approximately $15/ft of collection surface.[26] This price does not include tracking equipment, support structure or installation. Of course, the cost-effectiveness of this device would be optimum in an area having warm clear weather.

Scientists at Argonne National Laboratories, under the direction of Dr. William Schertz, are working on a device that utilizes a trough configuration and reflective surfaces to collect solar heat. The sides of the device (Figure 3.14) take on the shape of two different parabolic curves to focus incoming energy nearly uniformly over a narrow flat area at the bottom of the device. The original concept was developed by Dr. Roland Winston, a physicist at the University of Chicago's Enrico Fermi Institute, and is called a compound parabolic concentrator (CPG). The shape used is quite similar to the natural structure of the eye of the horseshoe crab, "one of the most efficient light-gathering structures known." [26] Interestingly enough, this similarity was not discovered until after the device had been pursued for some time.

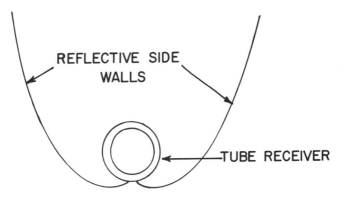

Figure 3.14. CPA reflector coupled to tubular evacuated receiver.

This configuration can focus light over a wider range of angles than can a conventional single parabolic reflector, minimizing the device's dependence on sun-tracking. Various concentration ratios can be designed into the device, up to a C-factor of 10 or so (a ratio that would, unfortunately, require full sun-tracking). If the C-factor were decreased to about 3, only one or two orientation adjustments would be necessary during each heating season. If the device is set to have a concentration ratio of 1.8, " it can be permanently set and the sun will fall within its acceptance angle throughout the year." [26]

A version of the device has been constructed of thin thermoformed plastic with a highly reflective aluminum coating. With the absorber tubes encased within glass tubes, effective operation could be achieved without the use of insulation, a significant material savings. Rough estimates of cost say that the unit could eventually be sold for about $10/ft^2 [26] and could operate at temperatures 150°F above ambient at efficiencies of about 50%.[12] 400-500°F output temperatures are possible under the right conditions.

A large number of organizations around the world are currently exploring parabolic cylindrical reflectors, paraboloidal dish reflectors, Fresnel configurations and other approaches to determine possible applications of high-quality solar heat to uses such as thermal process heat and the thermal generation of electricity. Arrangements both with and without tracking systems are being manufactured at present. It appears that high-quality solar heat (300-1000°F)[27] should be cost-effectively available within the next generation.

CHAPTER 4

THERMAL SOLAR ENERGY APPLICATIONS

INTRODUCTION

We have discussed the historical aspects of solar energy, its availability at ground level and the devices used to collect it's thermal potential. While interesting, this information is of little value without some idea of how to apply solar heat to specific purposes. This is the purpose of this chapter—to describe in overview fashion the ways in which solar thermal energy can actually be applied to satisfy specific needs.

It should be stated right at the beginning that this book deals with "active" solar equipment rather than "passive" systems. Many definitions of these two terms have been adopted by involved individuals and organizations, and no two are identical. The main ideas represented by the terms are essentially agreed on by all concerned, however, and boil down to the following: *"active" systems require the physical circulation of heat transfer fluids by means of pumps and/or blowers (which require outside energy to operate); "passive" systems rely upon natural convective, conductive and radiation principles to distribute heat, with no outside energy requirements.* In most cases, passive systems involve designs in which the building itself (or parts of it) serves to collect the sunlight. Many types of passive systems have been built and operated successfully, any many of the principles used should be applied to all new buildings to some degree, if only to reduce heating and cooling loads.

However, a detailed discussion of passive system technology is beyond the scope of this book, involving, as it does, a considerable number of architectural, climatological and human behavioral concepts. A quick lip service treatment of passive systems would not be worthy of this important subject, and, therefore, will not be included here. Good information on passive solar technology is available in almost any bookstore and from some government outlets as well, and should be consulted. All devices discussed here are used in an active mode.

Another point should be interjected here. Contrary to any exaggerated reports, solar energy systems are not capable of providing for the total annual heating or cooling load of a building. It is inevitable that periods of inclement weather will occur that are long enough to deplete the energy in any storage system, no matter how large. For this and reasons of cost, optimum solar systems are designed to deliver a certain increment (percentage) of the thermal energy required by a building during an average heating season. The specific optimum percentage varies with location, type and size of system and the purposes for which it is designed. The average optimum percentage of the combined demands of space heating and domestic hot water that a solar system can satisfy with present technology is about 60-85%.[28]

HEATING OF BUILDINGS

Just as the two basic working fluids used in flat-plate collectors are water and air, so too are the vast majority of buildings in this country heated with systems that utilize either hot water, steam or heated air to distribute thermal energy. In the case of designing and installing a solar heating system for an already existing building (this constitutes a solar system "retrofit"), the type of heating system already in place will determine the range of possible solar systems one can use. In other words, some types of solar systems will be more compatible with certain conventional heating systems and applications than others. Even if one is designing a new house or building and has the advantage of choosing both the solar and conventional (auxiliary) heating systems at the same time, the interrelationships between the two systems will be extremely important.

Depending on what specific functions are to be fulfilled, there are certain components basic to the successful operation of any solar heating system (Figure 4-1):

1. The collection element consists of an assembly of one of two of the types of devices described in the preceding chapter, mounted at a suitable orientation on or above the roof of the building (or, perhaps, on a ground location nearby). The function of this element is obvious.

2. The storage element serves to stockpile solar heat, so that it may be used indirectly at night or during relatively short periods of cloudy weather. Thus, it serves to "iron out" the peaks and valleys of solar energy availability, making the solar heating system more effective and reliable.

3. Distribution in a solar energy system has the same purpose as it does in any conventional heating system: to get the heat to where it is needed. For solar space heating, the distribution equipment used is indeed very similar to (although slightly larger than) that used in a "normal" system.

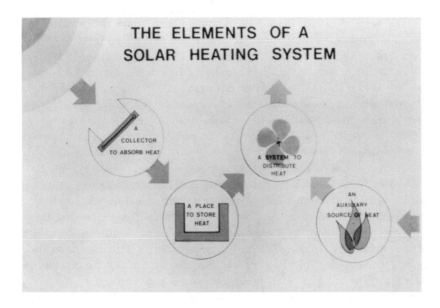

Figure 4-1. The four basic subsystems necessary to all solar energy space heating systems. (Drawing adapted from slide, courtesy of Sunworks, Div. Enthone, Inc.)

4. An auxiliary source of heat fills in for any failings the solar system may encounter. Even an optimally designed solar system cannot provide 100% of a building's annual heat requirements, since sunshine availability is at its lowest when the heat demand is at its highest. Storage extends the system's range, but cannot carry the load through an extremely cold or extended nonsunshine period. For these reasons, the building must be provided with a backup system of more conventional heating equipment, to heat the building when solar heat cannot.[29] Of course, this backup system must be sized to carry 100% of the building's heat demand.

Predictably, there is a large but finite number of possible ways in which these systems can be designed. As with any other situation, such as this, main characteristics must be used to classify the possibilities, to avoid saturating the issue with confusion. Since "most of the buildings in the United States are heated by delivery of warm air to the occupied space..."[30] this is where we will begin.

AIR SYSTEMS

The purpose of any space heating system is to attain an approximate equilibrium temperature in the occupied space that is comfortable to the occupants. Since every building loses heat to the atmosphere, this boils down to the heating system's responsibility to deliver heat to the rooms as fast as it is lost to the outside environment. This point of view makes it extremely simple to understand why upgrading a building's insulation would allow the installation of a smaller heating (and cooling) system. Once the building is as "energy-tight" as possible, then the heating system can be designed intelligently.

In most conventional air heating systems, a furnace (fired by oil, gas or coal) is used to heat the air to about 120-150° F or more. This heated air is then blown through an air ducting system and delivered to various rooms through outlets in the floor, ceiling or walls. A certain amount of cooling is experienced by the air between the furnace and the rooms, but this is usually negligible. The heated air then mixes with the air in the rooms, surrendering its heat until some equilibrium mixed temperature is reached. The amount of air needed to supply a certain building with the necessary heat, the size of duct required to carry that air and the appropriate capacity of blower for the system are all related to the temperature of the air to be delivered.

To utilize the air heated by an air-cooled, flat-plate collector array to space heat a building, the fact that only lower operating temperatures may be available at times must be allowed for. Although 120-150° F is not tremendously difficult to achieve on even a sunny winter day in the northern United States (return air to the collector from the rooms is usually at about 70° F, and air collectors can typically achieve temperature rises of 60-90° F), output temperatures would be significantly lower on a partly cloudy day. Thus, while the collectors are often receiving usable energy, the quality (temperature) of the energy is variable.

The variability cannot be eliminated, but can be minimized by storing the solar heat and releasing it at a controlled rate when needed. For storing the heat, the appropriate medium is as simple as a pile of rocks. A typical rock bed storage unit and its operation can be described as follows: a pile of washed gravel (1- to 2-in. diameter dense rock with a specific heat of approximately 0.2) is placed within a container. The walls of the container can be composed of just about anything of sufficient strength and low cost, such as concrete, brick, fiberboard, heavy-duty plywood, fiberglass-reinforced plastic, etc., the volume of rocks used to store heat should be sized at about 50 lb of rocks/ft^2 of collector area installed.[30]

The heated air usually is blown into the top of the rock bed and passes through the spaces between the individual pebbles, giving up heat to the pebble surfaces as it goes. Because very little heat is conducted from pebble to pebble (due to small areas of physical contact), insulation requirements to prevent heat losses from the storage bed are not very stringent, and 2 or 3 in. of fiberglass batting or its equivalent should be sufficient. For the same reason, it is fairly easy to attain good temperature stratification in the bed (some areas would be quite hot, others substantially cooler). Temperature stratification is desirable because it allows hot air (at the top of the rock bed) to be delivered to the rooms and also allows cooler air (at the bottom of the bed) to be returned to the collectors, maximizing efficiency.

The heated air is blown to the occupied spaces through an appropriate ducting system, and the ducting can either be routed through a conventional (auxiliary) heat source (furnace), or the auxiliary heat can be added with an "in-line" heater (electric resistance heating coils, water-to-air heat exchanger, etc). This allows the one ducting system to serve the house whether one or both of the heat sources are being employed. Figure 4-2 shows one way in which these components can be assembled to heat the house. Most systems are arranged such that heated air can be routed directly from the collectors to the rooms (bypassing storage), from the collectors to storage only (during periods of low heat demand), from storage to the rooms (with or without auxiliary heat), or from the auxiliary heater to the rooms (when no sunshine is available and storage is depleted). This adaptability allows for maximum use of the solar heat that has been collected.

HYDRONIC SYSTEMS

Operating temperatures during the winter for liquid-cooled collectors seldom exceed 120° F or so for any length of time, normally average about 85-105° F[31] and can vary widely, of course. A problem exists here, as most conventional baseboard heating systems are deisgned to operate with water circulating at about 180° F or more. Obviously, a much larger heat exchange surface (four or five times as great) would be necessary to heat as effectively with 100° F water, greatly increasing the heating system cost. The lower the operating temperature, the worse this situation becomes. This relatively low temperature water can be circulated to radiant floor slabs, should that kind of system be chosen, but radiant systems are not widespread, and have cost and operational problems of their own.

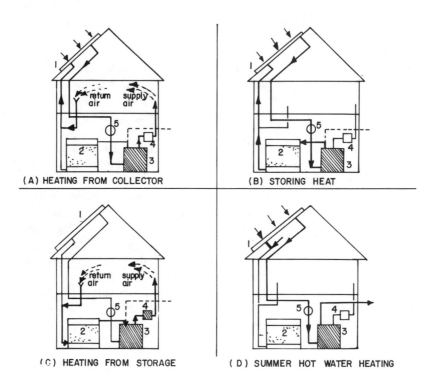

Figure 4-2. There are various ways in which the output from air-cooled collectors can be integrated into a solar space heating system. The one shown above is typical. (Reprinted with permission from ASHRAE *Applications of Solar Energy for Heating and Cooling of Buildings.*)

A water-to-air heat exchanger can be used to transfer the solar heat to air in a ducting system, and provides a workable method of using the liquid output of the collectors. It would obviously be helpful if the building's auxiliary system were of the air circulation type.

Appropriately, water is used as the storage medium in these kinds of systems. For a given amount of heat storage, rocks generally require about "3.5 times more volume than water storage..."[31] giving water a decided advantage insofar as space requirements are concerned. For sizing purposes, the most cost-effective water thermal storage quantity to use is between one and two gallons of storage water per square foot of collector surface.[30] Storage volumes much less than this would be unable to store sufficient heat to carry the building through an entire cold night, and volumes much greater than the two-gallon figure would be too large for the collectors to heat effectively.

Since water is excellent at transferring heat, the storage tank (constructed of concrete with a waterproof liner, various types of steel, fiberglass-reinforced plastic, etc.) must be covered with ample insulation materials to prevent wasteful heat loss. Approximately 6-8 in. of fiberglass batting or its equivalent are recommended for this purpose.[12]

As mentioned in Chapter 3, areas that experience freezing temperatures necessitate that liquid-cooled collectors are either drained when not in use or employ an antifreeze-water solution as the coolant. If water is used in a draindown system with a steel storage tank, the water from the collectors should pass through a heat exchange coil and transfer heat to the storage water without mixing with it. This minimizes any corrosive effects that would be experienced between the oxygenated water and the steel tank.[32] If an antifreeze solution is used as the heat transfer fluid, two different reasons arise to encourage the use of a heat exchange coil. First, it would be extremely expensive (if not foolish) to provide both the collector circuit and the much larger storage volume with antifreeze protection. Secondly, when the solar system also is used to heat domestic water (as is very often the case), any significant chance of contamination of the drinking water with antifreeze chemicals (often toxic) must be avoided. There are, predictably, many ways to exchange heat between collector output, storage and the energy load, one of which is shown in Figure 4-3.

CONTROLS

Insofar as the instrumentation of solar heating systems is concerned, its basic function is to command the system to use a maximum percentage of collected solar heat in a way to minimize the auxiliary input needed to do so (electricity for pumps, blowers, motorized dampers or valves, etc.). In most residential systems, temperature sensors are installed at the output of a representative collector panel, in the storage tank and in one or more of the heated rooms (thermostats). A central controller circuit receives these inputs and relays commands to the appropriate pump, blower or auxiliary heat source. For example, if the sensor on the collector registers sufficient temperature to indicate that useful energy can be provided by the array, but the thermostat(s) do not call for heat, the controller will turn on the pump or blower to circulate the transfer fluid. It will also open or close the right valves or dampers to route that heated fluid to storage.

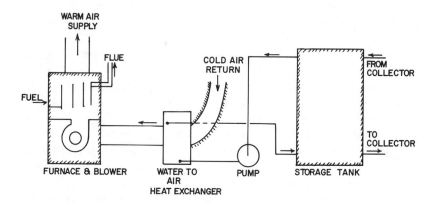

Figure 4-3. A typical solar space heating system, using solar heated storage water to heat the air flowing through a ducting system. In this system, auxiliary energy is supplied by a warm air furnace when necessary. (Reprinted with permission from ASHRAE *Applications of Solar Energy for Heating and Cooling of Buildings.*)

There is a wide range of difference in the complexity and sophistication of various control systems, but they are usually (or should be) the minimum that the system would need to operate properly. A control system is simply a rudimentary brain and nervous system for your solar equipment, similar to the one possessed by conventional heating systems of today.

COOLING OF BUILDINGS

The most fully investigated way in which the sun's heat can be used to effectively cool the living space in a building is to use it to drive an "absorption cooling cycle." In effect, this type of system accomplishes cooling by boiling an appropriate liquid. This seeming contradiction in terms can be cleared away by taking advantage of a paraphrased version of a beautifully lucid explanation originally offered by George Daniels.[33]

The illustrative case in point consists of a pot of water being heated on a stove. As the burner transfers its heat into the water, the water temperature rises until boiling initiates. After this point is reached (about 212° F at sea level), water temperature refuses to rise any further, regardless of the continuous addition of heat from the burner. This fact becomes a puzzlement (since the heat must be having some kind of an effect on the water), until it is explained that the heat energy is being utilized to turn part of the water into steam (the vaporous or gaseous form of water). The energy is serving to "change the state" of the water. The quantity of heat necessary to change a certain amount of 212° F water into steam is called its "heat of vaporization." This explains why skin feels cold after alcohol has been rubbed on it. The evaporating liquid obtains its heat of vaporization from the heat in skin. The discussion thus far is interesting, but must now be applied to understanding the absorption cooling process.

The major portion of the burner's heat is being absorbed and carried away by the rising steam. The burner, then, would be a great deal hotter if this were not occuring. In effect, the steam is actually cooling the burner, since cooling is a heat removal process. However, since we think of cool as being cool to the touch, let's talk now of more appropriate temperatures. Many fluids are available that boil at much lower temperatures than water. These fluids, of course, also absorb their respective heats of vaporization upon boiling. If the water in the pot were replaced with one of these fluids, the burner would not be necessary, so let's turn it off. The room temperature air surrounding the pot would supply sufficient heat to boil off the fluid, with the net result of a vaporized fluid and some cooled room air. The fluids mentioned above are generally called "refrigerants" and have boiling points extending substantially below zero.

To save the refrigerant for later reuse, the absorptive cooling systems used with solar heat are of the closed-loop type and generally operate as follows. As shown in Figure 4-4, heat from the solar system is used to heat up a strong refrigerant-water solution (the refrigerant is said to be "absorbed" into the water; hence, the process name) in an enclosed chamber (the "regenerator"). The refrigerant, being more volatile (having a lower boiling point) than the water (called the "absorbent" in this case), boils out of and separates from the water. The regrigerant condenses back to a liquid in the "condenser," is passed through an expansion valve (to dissipate heat) and is circulated to a heat exchange coil (the "cooling coil") located somewhere within the living space. This coil is surrounded by circulating room air, whose heat is transferred to the now cooler refrigerant, causing it to boil once again. The cooled air is blown back to the rooms, and the refrigerant vapor is circulated to a chamber (the "absorber") in which it is condensed and reabsorbed into the water. The heat that the refrigerant has taken out of the room air is released to either cooling water or to the outside air. A pump is used to return the remixed solution to the regenerator, completing the cycle.[12,33,34] This is an extremely simplified version of what actually takes place, but is enough to give an idea of the process.

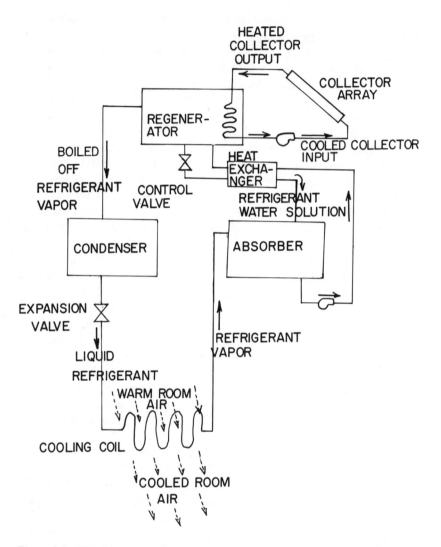

Figure 4-4. The diagram above is based on the lithium bromide absorption cooling cycle. An ammonia cooling cycle would be somewhat more complex.

In most cooling systems of this genre, the refrigerant mixtures used are lithium bromide and water (LiBr-H_2O) or ammonia and water (NH_3-H_2O).[35] Although some operational difficulties have been experienced with LiBr systems, it has a very low toxicity, as opposed to the significant toxicity of NH_3,[36] a very important difference between the two. An

NH_3-H_2O system also requires higher pressures and pumping horsepower, as well as a more complex system to separate the ammonia and water when necessary in the cycle.[35] For these reasons, LiBr-H_2O systems are the preferred choice at this point.

The solar heat input temperatures required to operate presently available LiBr-H_2O absorption equipment run around 240-300° F.[12,34] Operation at lower effectiveness is possible with input temperatures of about 180° F, however. It should be remembered here that, although flat-plate and even evacuated-tube type collectors are hardpressed to supply these temperatures consistently, relatively small auxiliary heat inputs will allow the cooling system to operate at full capacity. Another point to note here is that current LiBr cooling equipment was designed with quality heat units in mind (gas-fired, electrical resistance, etc.) and further research can very probably reduce the necessary input temperatures in time. Most of this equipment is also geared toward better compatability with hot water inputs than hot air input.[34] This too will change.

The controls needed for operation of the space cooling system are essentially identical in character to those needed with solar space heating systems, although different temperature set points will be used. Storage will serve the same purpose as previously outlined, that of storing thermal energy for use when needed (in this case to drive the absorption system). Advantage can be taken of these similarities.

Under favorable conditions, such as (ideally) a building with similarly sized annual heating and cooling requirements, solar heating and cooling capabilities should both be installed. This arrangement would allow the maximum total utilization of solar equipment all year long, and will show the largest total savings. A house with this kind of capability is shown in Figure 4-5.

DOMESTIC WATER HEATING

The simplest, and, presently, the most cost-effective way in which to use solar heat is to apply it towards producing domestic hot water (DHW). The cost-effectiveness is brought about predominantly by the essentially uniform year-round nature of hot water demand, allowing the system to be fully used at all times when sunshine is available. Hot water demands are possible to predict with a good deal of accuracy, using the same techniques used for years by designers of conventional DHW systems. The uniformity of demand and the fundamental objective of heating a tank of water cause DHW systems to be much simpler than space heating or cooling systems. These factors have caused solar DHW systems to receive the most engineering attention of any thus far, and "solar water heaters currently marketed are reliable items of household equipment."[37]

Figure 4-5. This "house of the future" was built at NASA's Langley Research Center in Hampton, Virginia. It is heated and cooled primarily with solar space conditioning systems. (Courtesy of NASA).

Solar water heating units were first introduced in this country in the 1930s, as discussed in Chapter 1. Similar systems are widely used in Japan, Australia and Israel.[38] The large number of these are designed to operate in a passive mode, by using natural convection (the "thermosiphon" effect) and thus will not be discussed here. Much information on these systems is available in the literature.[12,14,27,37] We are discussing pumped systems.

The collectors used for these systems all have the same design objectives of the collectors used for heating and cooling, and indeed are the same. The tilt angle used to mount collectors for this application is approximately equal to the angle of the local latitude, as mentioned earlier. While a properly sized solar DHW system can supply essentially all of the summer demand, auxiliary energy input should be provided for effective year-round DHW demand satisfaction.

Figure 4-6 shows a typical solar DHW system diagram. Solar heat is deposited in the storage tank (this would be the main storage tank if a more comprehensive solar system is installed). Storage water is fed to a second heat exchanger in the "pre-heat" tank. This tank's water is withdrawn and fed to the conventional DHW tank when hot water is used.

The main storage, as before, equalizes solar availability. Fluid flowing through the DHW heat exchanger allows the water in the preheat tank to heat up to as high a temperature as the stored solar energy can produce. The preheat tank gives the dwelling a greater DHW storage capacity than that provided by the conventional DHW heater alone. It also gives the water within it a longer period of time during which it can abosrb solar stored heat, lessening the instantaneous heating load on the conventional DHW heating element during demand periods. If the preheat tank were left out of the system, DHW storage capacity would be greatly reduced, and less of the total DHW energy demand would be taken by the solar system.

The name of the preheat tank illustrates the real averaged-out function of the solar DHW system. Over much of the working life of the system, solar heat is used to preheat the water, greatly lessening the energy demand upon the conventional DHW heater. This is the dominant function of the solar DHW equipment, and what makes it able to pay for itself.

SOLAR IRRIGATION

Approximately 260 trillion Btu are used annually by the agriculture industry in this country to irrigate more than 35 million acres of cropland.[39] In Arizona, California, Texas and New Mexico, there are currently more than 160,000 irrigation wells in use. The pumps for these wells

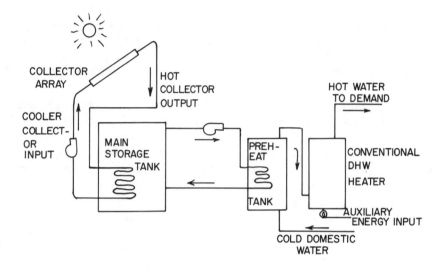

Figure 4-6. A typical solar DHW system, incorporating a preheat tank to maximize system effectiveness.

are powered by natural gas. Expected increases in natural gas prices may cause many of the wells to be shut down due to a loss of cost-effectiveness. Solar power can very probably be used to supplant natural gas and the other conventional energy inputs (electricity, diesel fuel, LPG and gasoline) for irrigation purposes, thereby averting a major food production crisis.

Solar power for irrigation has been provided experimentally by both photovoltaic (see Chapter 5) and thermal means. Basically, the thermal approach is designed around the operation of a solar heat engine. The engine is subsequently the thermal-to-mechanical energy converter, and is used to drive the irrigation pump.

Two such systems have been constructed on an experimental basis. One project is located at the Northwestern Mutual Life Company's Gila Bend Ranch near Phoenix, Arizona. It was funded by Northwestern at a cost of some $500,000. The project's solar pump system was developed by personnel at the Batelle Memorial Institute. The high-quality (high temperature) heat required for efficient operation of the heat engine dictated the necessity for solar concentration. Approximately 5,500 ft² of parabolic trough concentrators (focal length 36 in.) are therefore assembled into a nine-row collector field of 80-ft-long units. The trough array is, of course, endowed with full sun-tracking ability, and the troughs can be inverted to protect the reflective surfaces from damage by hail or stagnant conditions. An installation photo is shown in Figure 4-7.

The collectors output solar-heated water at 300° F, which is used to vaporize a refrigerant (Freon 113) in a closed-loop system (the principles involved here are similar to those used with the cooling cycle explained earlier). The gaseous Freon then drives a Rankine-type turbine (the solar heat engine), which drives a 50-hp pump capable of delivering a peak 10,000 gpm through a life of about 14 ft. No heat storage facilities are used at this installation. Batelle feels that the pump could be scaled up to 200-250 hp.[40]

Another project has been built on the Torrance County Land and Livestock Company farm near Willard, New Mexico. This experiment was funded jointly by ERDA and the State of New Mexico and is being conducted by Sandia Laboratories. The collector field at this site presently incorporates 6,720 ft^2 of parabolic trough concentration surface into 14 rows of 8 modules each. Figure 4-8 indicates how the system operates.

The Rankine-cycle engine in this system is indirectly driven by heat delivered by a hydrocarbon heat transfer fluid (Exxon's Caloria HT43) that has been solar heated to 420° F.[41] A mixing tank is also utilized by the system to control the input and output temperatures of the concentrator fluid. Freon 113 is used in this system (as in the Arizona project) as the working fluid. The engine delivers 25 shaft horsepower to the pump, which will deliver 880 gpm from a well 75 ft deep. A plastic-lined pond will be used to store as much as 4.5 ac-ft of water to supply the irrigation demand. This system uses a large thermal storage tank to equalize power availability.

It has been estimated that the system (when not used to drive the pump) could generate 10 million Btu of heat per day, or be used to generate 200 kWh/day of electricity,[42] suggesting uses beyond the irrigation season.

Results from both of these projects have been substantially encouraging. Further research and development of the concepts and hardware involved could permanently diminish the possibility of future shortfalls of irrigation water supply in the western United States.

THERMAL GENERATION OF ELECTRICITY

The use of large arrays of parabolic trough concentrators of the sun-tracking (just discussed) and fixed-mirror types (in which only the heat exchange pipe is moved throughout the day, and the reflective surface is stationary) to generate electricity has been investigated on paper. While technically feasible, significant thermal and pumping power losses would be introduced by the extensive matrix of conduits necessary to gather

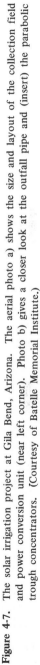

Figure 4-7. The solar irrigation project at Gila Bend, Arizona. The aerial photo a) shows the size and layout of the collection field and power conversion unit (near left corner). Photo b) gives a closer look at the outfall pipe and (insert) the parabolic trough concentrators. (Courtesy of Batelle Memorial Institute.)

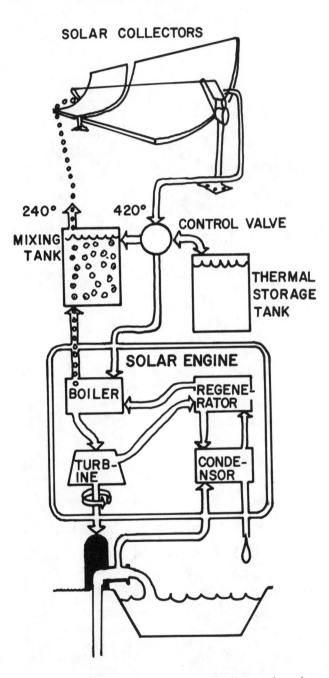

Figure 4-8. A solar powered irrigation system. This diagram shows how solar heat can be converted to mechanical energy for pumping applications.

and distribute the heated fluids.[43] It has been estimated that this type of energy transmission would be significantly more expensive than optical transmission.

The optical transmission approach is the main idea behind the power tower concept mentioned in Chapter 1. In essence, a large field of individually oriented mirrors focuses its cumulative energy onto a receiver situated at the top of a tower standing in the midst of the mirror field. The concentrated thermal energy received at the top of the tower is used to convert water to superheated steam. The steam drives a turbine/generator, resulting in an output of electrical power. Figure 4-9 gives an idea of what such a system might look like. The concept is being developed as part of the U.S. Solar Energy Conversion Program. Sandia Laboratories has been designated technical manager of this portion of the program, and has scheduled a course of development that should see construction of a 5-MW test facility (this should be in operation in 1978), two 10-MW pilot plants (the first should begin operations in 1980), and two 50-100-MW demonstration plants (the first is scheduled to be online in 1985). The culmination of the program should see commercial plant projects of 100-300-MW capacity being initiated in 1981 and 1983, with online operation projected to occur "after 1985."[43]

The individual sun-tracking mirrors are called "heliostats" and would cover approximately 30-40% of the total land surface bounded by the heliostat array (this "spread out" arrangement is necessary to avoid shadowing and transmission blockage effects between adjacent mirrors). The four different heliostat concepts developed during the initial investigations are shown in Figures 4-10 and 4-11. For a 10-MW pilot plant approximately 2,000 heliostats would be necessary (each having a flat reflective surface of about 430 ft^2) on 100 acres of land.[44] The tower near the center of the field would stand about 330 to 460 feet tall.

The heliostats will represent about 30-50% of the plant's total cost. The heliostat array would behave as if it were a huge adjustable Fresnel reflector, and could achieve concentration factors as high as 1,000,[43] allowing the system to operate at fairly high temperatures and efficiencies.

The use of water/steam as the circulating coolant appears to be the most promising at this stage (representing the least technical risk), although other coolant scenarios may look better as development progresses, such as liquid metals, hydrocarbon heat transfer fluids or molten salts. Designs put forth thus far are based on thermal storage capabilities of about six hours, which would allow for plant operation during most high-demand periods. This storage may be accomplished, depending upon the type of fluid used, by rock beds filled with fluid, a two-stage thermal approach (with separate storage tank for "hot" and "cool" fluids) or something

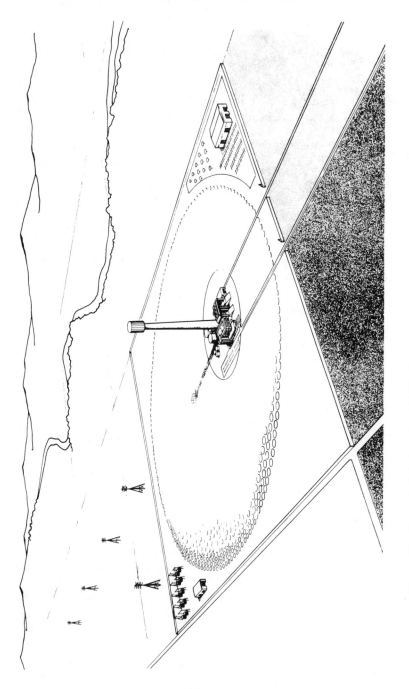

Figure 4-9. Conceptual design of a 10-MW pilot plant. (Drawing courtesy of Sandia Laboratories.)

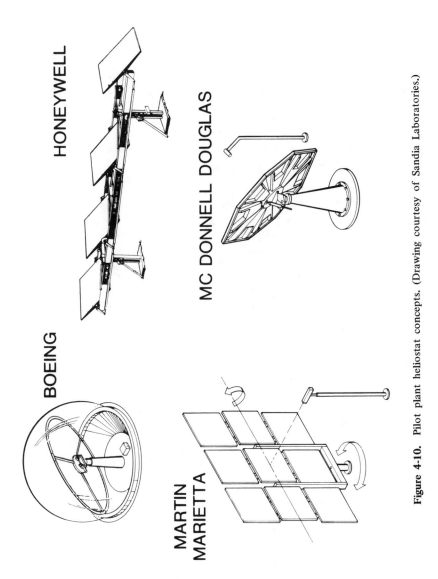

Figure 4-10. Pilot plant heliostat concepts. (Drawing courtesy of Sandia Laboratories.)

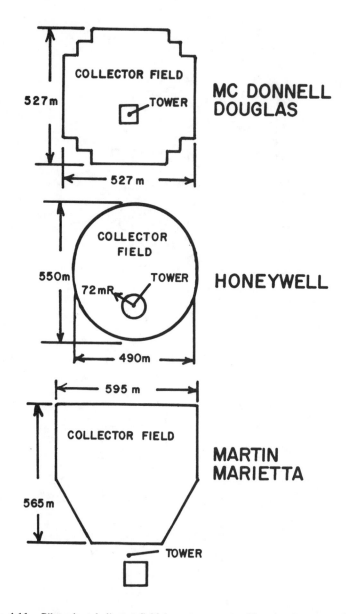

Figure 4-11. Pilot plant heliostat field layout concepts. (Drawing courtesy of Sandia Laboratories.)

not yet proposed. The eventual method used here depends on the rest of the system, costs involved and the load requirements of the local power grid.

After the turbine/generator is driven, the power conditioning equipment used to tailor the output before transmission will be quite similar to that used at present in conventional fuel-fired power plants. The conceptual operation of the system is summarized in Figure 4-12. On the whole, this type of solar power generation is closer to economic viability than many of the other solar power concepts currently under investigation, with the concept being "no more than a factor of 10..." away from cost-effectiveness.[43]

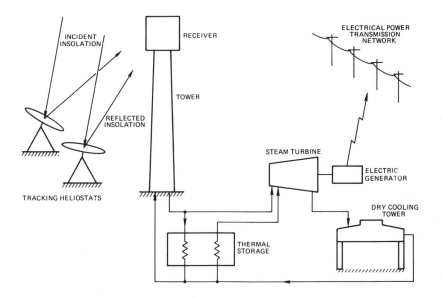

Figure 4-12. Solar central receiver thermal power system. (Drawing courtesy of Sandia Laboratories.)

CHAPTER 5

PHOTOVOLTAIC GENERATION
OF ELECTRICITY

INTRODUCTION

From an electrical point of view, the various solid materials that exist in nature or are synthesized by man exhibit a large difference in how easily they will conduct an electric current (flow of electrons through the material). Some are quite excellent at it, such as pure silver, gold or copper, and are classed as "conductors." On the other end of this spectrum, materials such as fused quartz have very low conductivities. These materials fall into a category known as "insulators." Materials that fall into the intermediate range of conductivities are known as "semiconductors." Without delving into the subject too deeply, a nutshell explanation of the structures of these materials would be helpful in understanding this chapter.

All atoms possess a certain number of electrons. These electrons are said to orbit about the atom's nucleus (center). Depending upon the number of electrons the material possesses, the orbiting electrons fill up a certain number of different concentric energy orbits or "shells" (each shell corresponds to a specific level of energy). As the number of electrons increases (with progressively more complex materials), each successive energy level is filled with the proper number of electrons, in a direction outward from the nucleus. The outer energy shell of a material will be either empty, partially or completely filled. It is this "distribution of electrons in the outermost or highest energy bands (that) determines most of the electrical and thermal properties of a material."[45]

The good electrical insulators are those materials whose outer energy bands are completely empty or filled, and an externally applied field of energy cannot easily change the energy state of the atom's outer electrons

81

(which would cause an electric current flow). Insulators also possess a fairly wide energy gap ("forbidden energy gap") across which the electrons would have to jump to support an electric current. Most metallic crystals have partially filled outermost energy bands, and can thus easily support a current flow.

Semiconductors are structurally similar to insulators, but have a much narrower forbidden energy gap. This allows semiconductors to support a current flow easily, after a certain minimum amount of energy input has been supplied (this energy must be of the correct frequency for the particular semiconductor being used). In the "dark," these same materials are good insulators.

PHOTOVOLTAIC EFFECT

In 1954 both RCA and Bell Telephone Laboratories achieved conversion efficiencies of 6%, using the sun as the energy source and cadmium sulfide and silicon as the "solar cell" semiconductor materials. The most widely explored solar cell material thus far has been silicon crystals of the "p-n junction" type. We will begin here.

A crystal of pure silicon is theoretically a perfect insulator, in that "a substantial amount of energy is needed to break the bond between an electron and the two atoms it binds together so that the electron can become available for conducting electricity."[46] When an electron is freed from its usual place in an atomic structure, it leaves a "hole" in its wake, which can easily accept another electron freed from a neighboring atom. It is this migration of holes and electrons that constitutes an electric current. In the real world, at any temperature above absolute zero, ambient thermal energy is sufficient to break some electrons free from a pure silicon atomic structure, allowing the material to conduct small amounts of electricity. This, of course, creates mobile electrons and holes in the material. If an external voltage is applied to the material, these electrons will travel to the positive potential, and the holes will travel towards the negative region. Augmenting this conduction ability is one of the main technical goals in fabricating solar cells.

The p-n junction arrangement is, so far, the most successful technique used for this augmentation. When pure silicon is "doped" (contaminated) with a small amount of, say, phosphorus, the basic silicon structure is maintained. However, the material will now have an excess of negatively charged electrons (from the phosphorus)—more than it needs to bond the atoms together. This material can readily accept more electrons, and is called an n-type semiconductor.

If the basic silicon is, instead, doped with an element such as boron, a hole-rich material results. This constitutes a p-type semiconductor, having an affinity for additional positively charged holes. A p-n junction silicon solar cell consists basically of a layer of both types of semiconductor materials. The n-type layer is situated on the side of the cell that will be exposed to sunlight, and is thin enough so that light can penetrate through it to the immediate vicinity of the p-n junction (where the two materials are bonded together). The energy contained within the sunlight will create free electrons in the n-type material and holes within the p-type material. This condition grows until "there is a voltage built up within the crystal sufficient to push any further electrons back to the p-type layer."[46] This creates an electric current. This behavior is roughly analogous to the flow of heat through a wall. The heat flow becomes significant when the difference between the temperatures on each side of the wall is sufficient to drive heat from one side to the other fairly quickly. Similarly, the use of the semiconductor doping technique creates two materials of different potentials. Applied energy increases this potential difference until electron flow occurs.

As a result, some of the incoming light is thus converted directly into electricity. The amount of electricity thus produced, divided by the total input light energy, is known as the conversion efficiency. This electricity can then, of course, be used to supply power to an external load. The basic configuration of a p-n junction solar cell is illustrated in Figure 5.1.

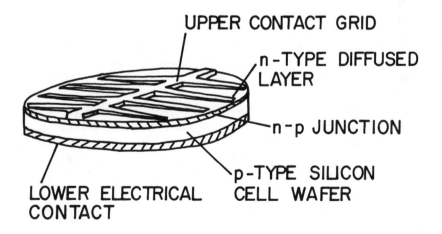

Figure 5.1. The basic configuration of a p-n junction photovoltaic cell. (The thicknesses are exaggerated for clarity.)

PHOTOVOLTAIC DEVICES

The Silicon p-n Junction Cell

These devices are composed, as stated earlier, of high-purity, single-crystal silicon wafers that have been precisely doped with trace elements. The electrical potential difference that is induced in these wafers by exposure to sunlight is trapped into by applying electrical contacts to both sides of the cell. The contacts (current collectors) on the "sun side" of the cells (the n-type side) are normally arranged in a "finger-like grid pattern designed to maximize current gathering while minimizing shadowing of the surface. An antireflective coating is also generally added to (the same) surface of the cell to minimize surface reflections and increase the energy absorbed."[47] Reflection and shadowing present no problems to the reverse side of the cell, so that current collector configuration is of much less significance.

As expected, the current available from these cells is directly proportional to the levels of insolation striking the surface. Typical output from these devices, at peak solar noon insolation, is approximately 10 W/ft^2 of cell area, with conversion efficiencies of 10-16%. Although the theoretical maximum efficiency for this type of cell is about 22%,[48] it is doubtful that more than 80% of that maximum will ever be achieved,[46] due to "real world" reasons.

The raw material for these cells, silicon dioxide (sand), is the most abundant element in the earth's crust, except for oxygen. It is effectively impossible to run out of this material, and its purchase price is almost absurdly low at about 0.25¢/lb.[49] The material is by no means hazardous, since it would eventually turn back into sand if destroyed by fire or breakage. There are no problems to be encountered with the availability or handling of this raw material.

A finished silicon solar cell, however, must be monocrystalline and possess a high degree of purity. While polycrystalline materials are able to perform in a semiconductor capacity, their performance is usually about 10% of that exhibited by cells of single crystal composition.[48] This one factor seriously complicates solar cell economics, since purification, crystal growth and shaping processes in use at present are extremely expensive.

The purification process basically entails high temperature melting of the sand and simultaneous reduction in the presence of hydrogen. This results in a very pure polycrystalline form of silicon. The cost of this refined silicon is a respectable $32/lb.[49] Purification, then, has singlehandedly increased material costs by more than four orders of magnitude.

The next steps are to reform this silicon into a single crystal and to section the crystal into individual wafers. There are currently two methodologies being explored to accomplish this. The first is known as the Czochralski growth method, first discovered in 1923, which results in single, perfect cylindrical crystals. A seed crystal of the material is dipped into a crucible containing a pool of molten semiconductor material (silicon in this case). This crystal is rotated and simultaneously pulled away from the molten pool, yielding a cylindrical crystal, typically three or four inches in diameter and several feet long when complete. The rates of rotation and "pull" must be controlled very precisely. A sketch of the process is presented here as Figure 5.2.

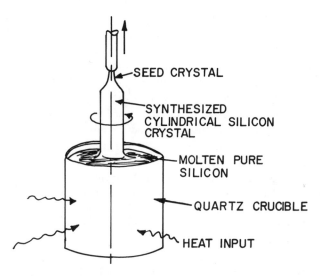

Figure 5.2. The Czochralski crystal growth process. This method is used in an environment containing an argon atmosphere. This inert gas helps to maintain the silicon's high purity.

To transform this cylinder into usable blank wafers, the crystal must be sliced carefully with a saw or abrasive disc. While this procedure is also very carefully controlled, material wastage is huge. In fact, for every pound of cut wafers, about 12 lb of crystal are lost as "sawdust."[49] The individual wafers are approximately 0.012 in. in thickness, the minimum thickness at which the wafer can handle further processing without breakage. They are, at this point, composed entirely of p-type semiconductor material. One side is subsequently doped by exposure to high-temperature phosphorus, forming a thin layer of n-type material. Electrical contacts are applied to the two surfaces, an antireflection coating is added to the n-type surface, and the entire cell is then sealed with a protective skin. Figure 5.3 shows cell manufacture.

Each of these cells now costs approximately $13 and can generate about two-thirds of a watt with peak sunshine intensity.[46] With typical efficiency for a cell of this type (15%), this translates to roughly $20,000 per peak kilowatt electric (peak kW) it can produce.[49] This is considerably better than the 1959 solar cell cost of $200,000/peak kW,[46] but the price is still exorbitantly high. The two main drawbacks with the Czochralski method are, therefore, high material wastage and the intensity of skilled labor necessary during fabrication. It has been estimated that the minimum future cost of this type of cell will be about $1,000/peak kW.[50] Thus, the requirements inherent in the process make it improbable that Czochralski silicon solar cells can ever be produced cost-effectively.

Another method, currently being developed at Mobil Tyco Laboratories by Mlavsky and Chalmers, offers a high degree of promise, however. The process is known as edge-defined, film-fed growth (EFG) and produces continuous ribbons of silicon crystal (of marginally less perfection than Czochralski crystals). An offshoct of extensive but unsuccessful development carried out during the last decade to produce the dendritic web type of continuous ribbon crystals,[51] EFG was originally developed for the growth of sapphire crystals for lamps.

The crystals produced by EFG are also drawn from a melt of pure silicon, but in a different way (Figure 5.4). The molten silicon is fed by capillary action between two halves of a die, by which action a ribbon is formed. "A film of liquid silicon at the top of the die feeds the growing crystal, whose cross section is defined by the shape of the top of the die."[46] The ribbon can be grown at about 1 in./min and is about 0.06 in. thick. The cooled-down crystal can then easily be cut into wafers 1 in. wide and 4 in. long.

It can easily be seen that much less waste will occur when cutting a ribbon into sections, as opposed to slicing a cylinder into many discs. The resultant rectangular cells can be fitted closer together than can an array of disc-shaped cells. Also, since crystal growth actually occurs at the top of the die, "thermal control of the crystal growth is sufficiently isolated from the silicon melt conditions such that multiple-ribbon growth from a common melt seems feasible."[51] It also looks to be quite feasible to continuously replenish the melt supply, breaking the process' last tie to batch operation. These factors make large-scale production (and the attendant reduction in costs) look quite achievable. Doping and final preparation of these cells would be similar to that used for the Czochralski cells.

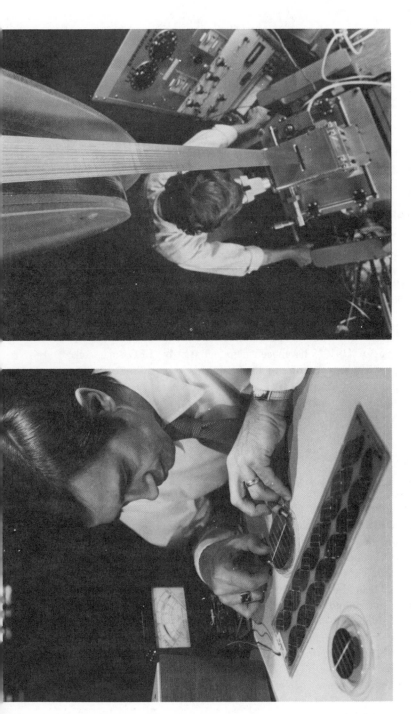

Figure 5.3. a) As shown in this picture, cells obtained from Czochralski crystals are disc shaped, and can subsequently be cut into sections to better fill the surface of a collector panel (note panel in photo). (Photo courtesy Batelle Columbus Laboratories); b) Silicon ribbon growing. (Courtesy Mobile Tyco Solar Energy Corp.)

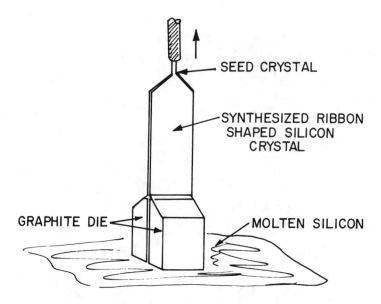

Figure 5.4. The EFG crystal growth process. This methodology shows great promise towards reducing, by mass production, the cost of silicon solar cells.

Efficiencies of 10-12% have been demonstrated by EFG solar cells, and the main problem seems to be in fabricating dies of materials that will not contaminate the crystals. One estimate states that the costs of these cells will shrink to about $250/kW.[50] A more conservative estimate (by Mobile Tyco) says that costs will be below $500/kW.[46] Reaching this latter figure would make solar cells economically competitive with conventional fuel burning power stations.

The Cadmium Sulfide p-n Junction Cell

This type of cell also constitutes a "sandwich"of p-type and n-type semiconductor materials, but there are considerable differences. Cadmium-sulfide (CdS), the n-type material used here, is deposited by evaporation onto a thin metal substrate such as zinc-copper foil, or a metallized plastic such as Kapton. The thickness of the resulting CdS film is about 0.0008 in.,[48] and the metal substrate serves as both negative electrode (current collector) and structural strengthener. The p-type material is applied by dipping the cell into a copper sulfide (Cu_2S) solution. This p-type layer adheres to the CdS by ion exchange (forming the p-n junction) and will be approximately 1% as thick as the CdS film. The positive current collector (a gold-plated copper grid) is then applied to the Cu_2S side of the cell by use of a photoetching procedure. The cell is finally sealed off by cementing a sheet of ultraviolet-resistant cover plastic (such as Mylar) onto the cell.[49] In cross section, the cell is similar in configuration to the silicon p-n cell, but the p-type and n-type layers are reversed (Figure 5.5). The surface area of a typical CdS cell is some 8.5 in^2., roughly the same as a silicon cell.

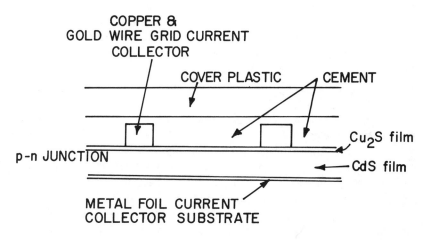

Figure 5.5. A cross section of a cadmium-sulfide solar cell. The thicknesses are exaggerated for illustrative purposes.

CdS cells have lower theoretical and achieved efficiencies, being approximately 8% and 5%, respectively.[49] They also possess a markedly greater tendency to degrade, due to moisture in the air and thermal cycling (this latter problem is caused by a "severe mismatch in the thermal expansion coefficients between the cadmium sulfide and the other layers making up the cell").[48]

The cost of CdS cells is significantly lower than silicon cells (since crystal growth is unnecessary), and it is estimated that costs can drop to as low as $600/kW within the next eight years.[49] The fabrication requirements seem compatible with mass production-scale operations, and CdS cells may well find a place in the future solar cell market despite their low efficiencies and instabilities. It must be remembered that the technology for thin-film cells is still in its embryonic stage.

Other Types

Amorphous (uncrystallized) silicon, as the name implies, has an extremely disordered internal structure. It is relatively nonreceptive to doping procedures, and predictably tailoring its optical and electrical behavior is fruitless with the usual semiconductor methodologies. In recent years, however, it has become evident that the addition of hydrogen to the amorphous silicon made it more receptive to doping, which, in turn, made it possible to fabricate photovoltaic devices with the material. RCA is working with a process that would use layers of undoped amorphous silicon, platinum, phosphorus-doped silicon and a stainless steel substrate to form a "Schottky barrier"-type of workable solar cell.[52]

Much more data must be generated on this type of device before its fate becomes evident. Present problems with low efficiencies and transient, long-term performance (gradual changing of the amorphous material's electrical properties) will probably be the major barriers to the cost-effectiveness of this approach.

During about the last five years, researchers from all over the world have begun to announce the exploration of an incredible number of various solar cell approaches, all aimed at producing an inexpensive, reliable, mass-producable solar photovoltaic cell. Thin-film research is being carried out with selenium, cadmium-telluride, cadmium-selenide, zinc-telluride, indium-phosphide, gallium-phosphide and others as the semiconductor material. Some of these materials are suitable for "heterojunction" cells (junctions composed of two different semiconductors.[53] IBM has fabricated an experimental gallium-arsenide cell that exhibited an efficiency of 22%.[54] Multiple-junction cells (those with stacks of two or more cells, each utilizing different parts of the solar spectrum) reportedly have theoretical efficiencies of up to 40%, and are being researched by Varian Corporation and Texas Instruments.[54]

It is patently impossible at this point to foresee which of these approaches will take the lead in the ultimate fruition of solar cells as widely accepted sources of energy. The fact that industry is becoming committed to the problem, however, brings that fruition closer and makes it much less uncertain.

Output Increases

Developing cheaper fabrication processes is not the only way to make solar cells cost-effective. As stated earlier, the output of solar cells is directly proportional to how much solar energy falls upon them. This suggests the use of a principle first discussed in this book in connection with thermal collection techniques...concentration of the sun's energy. Obviously, each cell in an array would put out much more electricity when receiving energy at a much higher rate.

Experiments with silicon solar cells have shown the concentration of sunlight will, indeed, increase solar cell output, if the cells are not allowed to get too hot. If they are allowed to overheat, cell performance and operating life suffer seriously, negating any advantage that concentration may have provided. Concentration ratios of as much as five can be used with the cells before the excess heat must be actively removed.[49] This kind of concentration can be supplied most effectively with the compound parabolic trough concentrators developed at Argonne National Laboratories (see Chapter 3). At this factor of concentration, only seasonally periodic tilt adjustment of the array would be necessary.

As the concentration ratio increases, full sun-tracking and solar cell heat removal become necessary. This fact eliminates the stationary solid-state reliability of the cells, since they formerly operated with no moving parts or circulating fluids whatsoever. If these additional complications can be tolerated, however, experiments have shown that silicon solar cells can operate in a system that exposes the cells to a solar flux of over 180 peak W/in^2. (a concentration factor of 300). The output of these externally cooled cells was more than 9.7 W/in^2., representing an output increase of well over two orders of magnitude.[55] The effect of this elevated output on the life span of the array has not yet been determined.

These cases represent only two of those investigated thus far with regard to concentrator/solar cell combinations. Many variants exist, and more are conceived as the months go by. Varian Corporation has built an experimental system operating with a concentration ratio of 1,735. The system operated at 19% efficiency, and output was reported to be over 150 W/in^2. of exposed cell area.[54] Gallium arsenide cells are reportedly more well suited to concentrated solar fluxes, being able to operate at temperatures as high as 390°F without serious losses in efficiency.[54]

The need to cool the cells result in the transfer of low-grade heat to the cooling water. This water can then be treated as if it had been output from a flat-plate collector. This kind of arrangement is being explored at Sandia Laboratories in New Mexico (Figure 5.6). Their system uses flat Fresnel lenses to concentrate sunlight 50 times, allowing the 135 silicon solar cells to produce 7.4 peak W each. The array is capable of full sun-tracking movement. The hot water leaving the array can be used for process hot water, domestic hot water, absorptive air conditioning or space heating demands.

This system can be considered representative of many of the concentrator/solar cell arrangements currently under development, and brings up the intriguing possibility of providing both electrical and thermal energy from one hybrid collector. This concept is known as "solar total energy," and is being actively investigated by Sandia Laboratories at its test facility in Albuquerque.[56] Hard operating data are currently being gathered and should be ready for presentation in the near future. It is

Figure 5.6. The Sandia photovoltaic concentrator system. The flat Fresnel lens concentrate sunlight onto the solar cells mounted beneath them (seen as white dots in this picture). In turn, coolant fluid is circulated below the cells to cool them off and to obtain solar thermal energy. (Photo courtesy Sandia Laboratories.)

expected that such systems will compete favorably with conventional fuel-fired systems, especially in small-scale applications.

ARRAY SYSTEMS FOR ELECTRICAL GENERATION

Terrestrial Generation

Not surprisingly, no MW-scale photovoltaic arrays have been constructed as yet. Before they can even be conceptualized with any kind of accuracy, experience gained thus far with numerous smaller-scale systems that have been built must be digested and expanded.

Until recently, the only market of any significance for photovoltaic devices has been for powering the orbiting and deeper-space excursion missions fabricated by space technology. As a result of the hundreds of space photovoltaic systems that have been and are being used to power the vast majority of the spacecraft launched by the United States and the Soviet Union, a sizable body of experience has been built up on the technology of small-scale solar cell usage. The goal of these developments has been directed toward reliability and size reduction, however, and, while successful, resulted in a technology of expensive devices.

This know-how is now being adapted for terrestrial uses in many areas and in many locations, with the emphasis now placed on cost reduction approaches. Since 1972, the Japan Maritime Safety Agency has used solar cell systems to provide power for 256 lighthouses along the coast and on the various islands of Japan.[57] That country's PROJECT SUNSHINE will actively investigate the cost reduction and subsequent utilization of photovoltaic cells until at least the year 2000.

In this country, NASA has been the major factor in terrestrial development, and is administering the major portions of the Energy Research and Development Administration's Photovoltaic Conversion Program. (The newly formed Department of Energy will doubtless absorb ERDA's responsibilities in this and other programs). The NASA Jet Propulsion Laboratory has been assigned responsibility for the Low-Cost Silicon Solar Array Project, whose objective is to reduce the costs of solar cell arrays to 5% of the present level, or lower, by 1986. The project is being conducted such that the major portion of actual research is being carried out by the private sector (industry and institutions of higher education).

A number of prototype devices generated by this effort will then be tested by the Lewis Research Center at its System Test Facility. The facility "currently has a test capacity of 10 kW (and) is being used to develop an understanding of how these arrays may be successfully integrated with direct current and alternating current inverters, controls, and so forth in effective power systems for a variety of uses."[58] Lewis has also been developing small photovoltaic systems for use with remote refrigeration systems, to power an experimental highway dust warning sign on an Arizona highway, and complete power systems for two U.S. Forest Lookouts (lookout stations) in northern California (each with a 300 peak W capacity), among others.

MITRE Corporation, a nonprofit research organization, installed a 1 peak kW-capacity solar cell array on the roof of its Washington facility in 1975 (Figure 5.7). After four man-years of effort, the experience gained pointed out the need for substantial development of available equipment, including collection and power conditioning equipment. The program helped to delineate some of the problems to be anticipated in working up to an actual large-scale system.[59] Degradation caused problems with the solar panels, such as delamination of the cover plate from the cell face (with moisture entrapment in the resulting space, causing corrosion), gas bubble formation and other environmentally caused factors.[60] This continuing study is a part, of course, of the U.S. Photovoltaic Conversion Program.

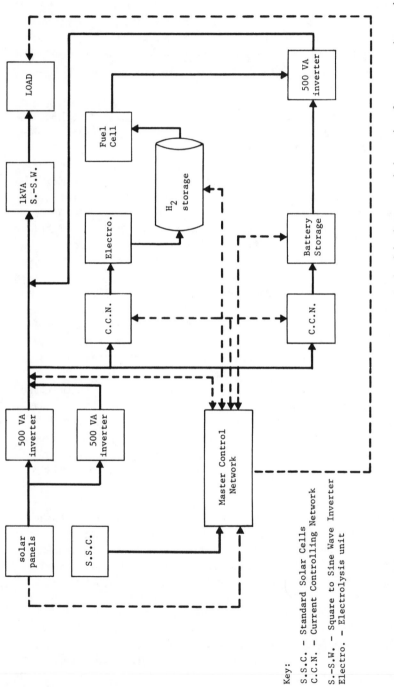

Figure 5.7: The initial configuration of the MITRE Solar Energy System. Lack of funds prevented the project from progressing much past this state.

Key:

S.S.C. – Standard Solar Cells
C.C.N. – Current Controlling Network

S.-S.W. – Square to Sine Wave Inverter
Electro. – Electrolysis unit

The largest photovoltaic project yet constructed in this country is a 25-peak kW system operating in Mead, Nebraska. The power output of the 100,000 silicon solar cells has been used for two purposes. Its original application was to power a 10-hp pump for 12 hr/day, to pump irrigation water to 80 acres of corn. After the growing season was complete, the output power was rerouted to operate a "natural air drying technique... being tested on the harvest corn with large circular fans powered by the photovoltaic array."[61] Two grain bins, with individual capacities of 6,000 bushels, are each equipped with a 5-hp fan that provides 1 ft^3/min of air per bushel of corn to be dried.

The system incorporates an 85-kWh capacity bank of lead and storage batteries to even out power variability. The output of the stored power is subsequently conditioned by three electrical inverters, which change the dc power into the 240-volt ac power used to drive the load (pump or fans). The project was funded under ERDA's Photovolatic Conversion Project in the amount of about $1 million. It is being conducted by MIT's Lincoln Laboratories and the University of Nebraska-Lincoln.

Based upon the information gathered thus far, the cells used operate with a conversion efficiency of approximately 10%. R.W. Matlin of Lincoln Laboratories has estimated that the eventual price of a similar system, when commercially available in the mid-1980s, will be $40,000 or less.[61]

The program intends to install a number of small and moderately sized array systems at various military and governmental installations within the next decade, in a series of stepped programs. The culmination of the development schedule is projected to occur in the mid to late 1980s, by which time cumulative demonstration capacity should be some 10 million peak watts of solar output.

Myriad studies will be undertaken to explore a wide variety of semiconductor devices, power conditioning systems, storage techniques (to spread out the solar array's base-line capability), concentrator scenarios and conceptual designs for small-, medium- and large-scale generating systems.[62]

Before these studies come to fruition, it would be an exercise in conjecture to try to predict which type of array and power-handling system arrangement will be the optimum for terrestrial use. If past advances in similar types of technologies are any indication, however, the concerted effort that is just beginning will fill the next 10 or 20 years with many pleasant surprises.

Space-Based Generation

As mentioned in the last section, a major drawback to the use of terrestrially based photovoltaic generation is the same problem that plagues

Figure 5.8. The SSPS concept: **a)** A solar satellite power station, as depicted in this artist's concept, is being studied under contract to NASA's Lewis Research Center. Such a station, placed at synchronous altitude, would convert the sun's radiation into electricity by large solar arrays. This energy would be transformed to microwaves and transmitted to earth where a receiving station would convert the incoming microwaves into dc electricity at a very high efficiency. The satellite concept is one

of several methods being explored as ways to generate large amounts (megawatts) of electric power for uses on earth; b) Solar Power Satellite. An artist's concept depicting one of the solar power satellite microwave transmission antennas being assembled at the outer edge of the structure. A solar power satellite would be located in a geosynchronous orbit some 36,000 miles above the earth. (Photos courtesy of NASA.)

all solar energy systems on earth: interruptions caused by bad weather and the day-night cycle. There is no location on earth where this problem can be avoided. It can only be minimized and allowed for, usually at great expense. However, a generating station placed in a location above the earth could be free from such interruptions, and also take advantage of the solar constant's full strength (by intercepting it before it reaches the atmosphere).

The Satellite Solar Power Station (SSPS) would be placed in a "geosynchronous" orbit, at an altitude of approximately 36,000 miles above the surface. At this altitude the station could remain directly above a fixed point on the surface, in the same way that communications satellites do now (hence, the term geosynchronous). The SSPS would be able to receive a total of 6-20 times as much solar energy as would be available to an identically sized collection area on the planetary surface.[63] Its continuous power input would be interrupted only near the seasonal equinoxes, "at which time the satellite will be shadowed by the Earth for a maximum of 72 minutes per day. Averaged over a year, this shadowing results in only a 1 percent reduction of the energy that would be available if the SSPS were continuously exposed to sunlight."[64] Also, the shadowing caused by this phenomenon would occur near midnight, at which time consumer power demand is quite low.

The SSPS, as presently conceived by NASA and others after several years of study, would consist of two, large, solar cell-filled collection surfaces of some 11 mi^2 each, supported by an immense superstructure designed for zero-gravity conditions[64] (Figure 5.8). These panels are designed to supply approximately 8,500 MW of electricity to a microwave transmitting antenna situated between them. The total mass of the station is estimated at about 40 million pounds, and yields a mass:earth received power ratio of only 8 lb/KW (this is extremely low compared to terrestrial systems—zero gravity has its structural advantages).

The microwave transmitter, after converting the input power to microwave radiation, beams that radiation to a receiving antenna at a specific earthly location. It is expected that 5,000 of the original 8,500 MW will be received at the surface. This received power will then be rectified back into dc or ac power, for delivery to the conventional power grid.

This type of process was demonstrated during the summer of 1975, at the NASA Venus antenna site of the Goldstone Tracking Station near Barstow, California (Figure 5.9.) The 85-ft diameter dish, normally used for communication with spacecraft, was positioned about one mile from the 25-ft-high set of receiver panels shown in the distance. Microwave power was beamed from the Venus dish to the receivers, in turn lighting the array of 17 lights on the tower. An average conversion efficiency of 82% was obtained during the experiment,[65] proving conceptual feasibility on a small scale.

Figure 5.9. In 1975 microwave power was transmitted from the Venus dish antenna in the foreground to the receiver tower in the background, one mile away. The
lights at the bottom of the tower are being powered by the converted microwave
radiation that the tower is receiving. This is NASA's Goldstone Tracking Station
near Barstow, California. Normally used to communicate wtih interplanetary spacecraft, the big 85-ft venus dish, as it is called, is now serving as a research tool to
study the problems of beaming converted solar energy from satellite in space, back
to earth where it can be reconverted to electricity. For the tests, the 85-ft antenna
represents an energy satellite. From its vantage point in space, it would collect and
convert sunlight energy into electricity. A mile away a 25-ft-high set of receiver
panels called the Rectenna, plays the role of a ground station. Made of aluminum,
the more than 4,500 small rectenna elements are T-shaped, and about 4 in. long.
They work like a TV antenna to gather and filter the microwave energy converting
it to ac or dc that could be fed directly to a utility. With the antenna zeroed-in
on the mile-away receiver panel, microwave power transmission can be ordered to
bring power up slowly in 25-kW increments. On a hillside beneath the receiver
tower, a bank of 17 lights begins flickering on, reaching full intensity as the power
increases. The lights go off and on as the big antenna tilts up and down and swings
right and left – now on the rectenna, now off. The results have been very promising
as Goldstone researchers successfully collected the microwave beams and converted
it to usable electricity with an 82% efficiency. (Photo courtesy of NASA.)

The receiving antenna in the scaled-up system would be approximately 6.2 miles in diameter "to achieve a reasonable power density within the microwave beam at the receiving antenna for efficient conversion of microwave into DC." [64] Incidentally, the antenna itself would be about 80% transparent to normal sunshine, raising the possibility of dual use of the land.

The effects of prolonged microwave radiation exposure on health and radio-frequency communications must be thoroughly investigated, although it has been estimated that the power density within 6 miles of the beam center would meet the lowest international standards for prolonged continuous microwave exposure. Waste heat produced at the receiver would be substantially less than that "released from energy production methods based on thermodynamic cycles" [64] and, in all probability, only natural air convection would be necessary to dissipate it.

The materials for the SSPS would either be transported to space from the surface of earth and assembled in orbit, or obtained as raw materials on the lunar surface, from whence it could be refined and processed at a permanent space manufacturing facility[66] (an engrossing concept in itself). Transportation costs from earth are expected to approximate between $20 and $60/lb delivered to low earth orbit. Each SSPS will require about 60-100 heavy-lift vehicle launches.[64] It is also estimated that as many as 112 SSPS installations could be placed into geosynchronous orbit above the U.S. mainland, supplying perhaps 40% of the country's total projected electrical demand in the year 2025, before that immediate region of space begins to get "crowded."[65] As is evident, the SSPS concept is unconventional, fascinating, possessed of an immense potential and certainly demands further serious investigation.

CHAPTER 6

ENERGY FROM THE WIND

INTRODUCTION

Of the solar energy that fails to penetrate our atmosphere, the vast majority is reflected back into space by dust particles and cloud cover. Very little actually goes towards heating the air. The main mechanism of atmospheric heating is, perhaps surprisingly, conductive and convective heat transfer from the sun-warmed land and oceans. As discussed in Chapter 2, terrestrial declination, solar angle of approach, local terrain structure and cloud cover prevent the earth's surface from being heated uniformly (causing temperature differences on the surface and in the atmospheric masses situated near them). These temperature differences cause atmospheric pressure differences as well, and the air attempts to equalize these peaks and valleys by circulating across the terrestrial surface in swirls and eddies. In a sense, the wind is the heat transfer mechanism to the earth's collector, and is the main agent behind the continuous interplay of natural processes we call weather. In this way, the dynamic heat equilibrium of earth is maintained.

Thus, it can be seen that wind energy is directly and totally caused by solar insolation and is, in fact, "solar energy once removed." [67] Wind energy is a kinetic entity, an energy source expressed as the physical movement of a fluid. Being a mechanical form of energy, it must be collected in a mechanical fashion, by devices that dynamically react to the moving air in an efficient and useful fashion, extracting as much of the wind's kinetic energy as practically feasible.

This brings us to a short discussion of wind energy itself, so that you can form a useful mental picture of just what it is we're dealing with. As you would expect, the amount of kinetic energy contained within moving air depends on its density (mass per unit volume) and its velocity. The effects of changing both of these parameters are very much as predicted

101

by your intuition. Much more kinetic energy is contained, for example, within a cubic foot of moving water than in a cubic foot of similarly moving air (in fact, air is approximately 800 times less dense than water). And the faster that moving volume of fluid is traveling when it collides with something (like the vanes of a windmill), the greater the amount of energy transferred. Velocity, put another way, allows a greater number of fluid molecules, each having greater kinetic energy than if slowed down, to transfer their energy to the windmill within a given time. The density of the atmosphere undergoes no tremendous changes near the earth's surface (at those elevations at which engineering feasibility forces us to collect wind energy), and, thus, remains essentially constant, eliminating density as a major variable that can be optimized. This leaves air velocity (and ways in which to harness it) as the area of concern.

As a result of extracting energy from the wind, the velocity of the air diminishes. If a windmill were 100% efficient, the air traveling through it would come to a dead stop, depleted of kinetic energy. Since it obviously cannot tolerate a mass of dead air accumulating immediately downwind of your windmill (preventing new air from passing through), it makes sense to assume that there is some theoretically maximum percentage of energy that can be extracted without causing a device to work against itself. The generally accepted maximum fraction is about 59.3%, a number obtained some 50 years ago as a result of the work of Betz.[68] If a wind machine were to extract this much of the wind's power, the wind passing through the blades would slow down to approximately one-third its original velocity. This machine would be said to be operating at 100% of its theoretical maximum efficiency, although it is gathering slightly less than 60% of the actual total energy available. When reading efficiency ratings on a wind machine, make sure you understand the basis of the efficiency given.

The power output of a windmill increases linearly as the area swept out by its blades increases, simply because that much more air is encountered. For machines having horizontal axes (discussed further along), this means that power output would be proportional to the second power of the blade diameter. Thus, doubling the blade diameter on a wind machine would (on paper) quadruple its power output (2 raised to the second power). Another such relationship, even more impressive than the first, states that obtainable power increases proportionately with the cube of the wind velocity. This means that a windmill's power input should increase eight times (2 to the third power) when the wind velocity doubles. You can see from this that changes in wind velocity can cause enormous differences in the power output of a wind machine. It is this desired constancy and magnitude of wind velocity, together with the behavior of certain materials and structures, that defines many of the problems associated with picking suitable sites for and the types of wind machines to be used to harness the wind.

The use of wind power has been explored to various degrees in many ways, by many peoples, for many centuries. Wind power has been used to power ocean-going vessels, drive carousels, produce music, pump water, grind grain and (more recently) to generate electricity. In fact, the term "windmill" refers specifically to the devices that used wind power to mill grain in various regions of the world and is inappropriate to any other application. Many names have surfaced recently to describe the other wind devices, such as "wind turbine," "wind-driven generator," "wind energy conversion system," etc. These new terms are many, varied, usually accurate and almost always unwieldy. Since it will probably be some time yet before one of the new terms becomes widely used, "windmill" will be used in this chapter as a catch-all term for wind-powered mechanisms (its present context). More descriptive titles will be used when appropriate.

By and large, the main application for modern windmills would be for the production of electricity, and it is towards this end that the discussions and descriptions in this chapter will be directed. As with solar thermal and photovoltaic conversion devices, there are many approaches taken to collect energy from the wind. Therefore, the time has come to divide the different approaches into types, to simplify the discussion. All windmills react to the wind by rotating around a central axis. These axes are situated either vertically or horizontally, or, to be more accurate, perpendicular or parallel to the direction of oncoming wind. This fact defines the two major categories of wind machines. Both are currently being investigated in various guises.

HORIZONTAL-AXIS MACHINES

Historically, greater experience has been gained with machines having rotational axes oriented parallel to the flow of wind. Hundreds of different designs have been generated within this subject of devices, some of which have not been in use for centuries. The Greek and Portuguese windmills with their triangular cloth sails, the northern European (predominantly Dutch) post mills, the smock mills used in northern and western Europe and in America, the more efficient (less aesthetic) American farm windmills, the Sailwing concept (first prototyped at Princeton University) the huge Smith-Putnam wind turbine operated in Vermont during World War II, the ERDA-NASA Model Zero wind turbine recently constructed in Ohio, the numerous small-scale manufactured units available (Figure 6.1), and the many new variations emerging (chalk rotor, Salter rotor, etc.) all belong to this class of devices.

Figure 6.1. The units above are but a few of the numerous smaller-scale units now available commercially. Due to the historically greater availability of cheap fossil fuel in this country, most of these commercial wind machines are currently produced in Europe and Australia. (Photo courtesy of Rockwell International.)

Windmill usage (for electric generation) in this country peaked during the 1930s,[69] at which time over 300 manufacturers existed throughout the world. Between 1880 and 1930, it has been estimated, "six million small windmills were built for farm use" [70] in this country. In the beginning, all were essentially used to pump water for irrigation. Some were used to drive simple stationary farm equipment (to grind grain, etc.). The windmills used in the United States during the mid-1800s produced a cumulative total of approximately "1.5 billion horsepower of work - equivalent to almost 12 million tons of coal,"[67] an impressive amount of mechanical power. As time rolled on, many new windmills were purchased, and old ones were converted for use as wind-driven generators. Indeed, wind-driven generators appeared to have a secure future (both electrical and aerodynamic technology was now being applied to the field), until the arrival of cheap widespread electricity with the Rural Electrification Administration and the availability of small inexpensive engines cut that future short. The temporary era of cheap fossil-derived energy had begun. Much wind power activity ceased; however, not all.

THE SMITH-PUTNAM WIND TURBINE

Palmer Coslett Putnam, a graduate of M.I.T., built a house on Cape Cod in 1934, and began looking into whether a windmill might supply his all-electric dwelling with power more cheaply than did the local utility. He soon found that small, commercially available windmills would be insufficient for his needs. Subsequent comprehensive study convinced him that the cheapest wind power scenario would consist of "the direct generation of alternating current by a very large, two-bladed, high-speed windmill, feeding (power) into the lines of an existing hydro (electric) system. Thus, existing hydro-storage would provide the capacity to tide over periods of no wind." [71] As time went on, Putnam's idea grew and developed, with the help of an amazing cadre of technical experts and businessmen.

The actual project was officially born in October of 1939, in the office of Thomas S. Knight, then commercial vice-president of the General Electric Company in Boston (GE agreed to develop and supply a synchronous generator to tie in the wind-generated power into the New England power grid.[72] An abbreviated list of the people involved includes: Dr. Vannevar Bush, Dr. Severre Petterssen and Professor John B. Wilbur, the Dean of Engineering, Head of the Department of Meteorology and Head of the Department of Civil and Sanitary Engineering, respectively, at M.I.T.; Beauchamp and Burwell Smith, both vice-presidents of the S. Morgan Smith Company (then the leading manufacturer of controllable-pitch hydraulic turbines); Albert Cree and Harold Durgins, president and vice-president (and chief engineer) of the Central Vermont Public Service Corporation; and Dr. Theodor von Karman, the head of the Aeronautical Laboratory at the California Institute of Technology.

The design and fabrication of the wind machine were severely rushed, due to the immediate threat of American involvement in World War II. As a result, the huge forgings for the machine were ordered "on the basis of approximate and fairly rough estimates of stresses," [72] as a full detailed design for the unit was not possible in the short time available. Grandpa's Knob, a 2,000-ft hill located somewhat west of Rutland, Vermont, was chosen as the site for the installation. Construction was started during the winter of 1940-1941 and finished in August 1941. The completed machine stood on a 110-ft tower and used two blades, each 11 ft wide, about 70 ft long and weighing approximately 8 tons. The entire turbine weighed about 250 tons and was arranged so that the blades were down-wind of the tower (Figure 6.2). The unit had a blade diameter of 175 ft, a hub height of 125 ft and a generator speed of 600 rpm. [71] An elevator was used to move personnel to the control room at the tower's summit.

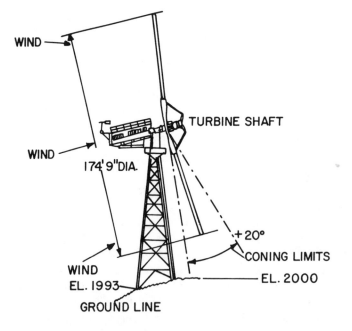

Figure 6.2. The Smith-Putnam wind turbine on Grandpa's Knob in Vermont. With its 1250-kW capacity the installation has maintained a 33-year record as the most powerful wind machine ever built by mankind.

Full-fledged operation commenced on October 19, 1941, after more than a month of operational preliminaries with the benefit of a gusty 25-mph wind coming from the northeast. Under these conditions, the turbine blades were adjusted (this refers to blade "pitch" primarily, the angle at which the blade surface presents itself to the wind) so that power output reached 700-kW. At this point, the output power was synchronously fed into the utility power grid, a first for modern technology.[71]

The wind turbine was rated at 1250-kW output, but generated outputs as high as 1500-kW during 70-mph winds. With the blades fully "feathered" (blade surfaces turned practically parallel to the wind) and locked, the machine was able to sustain 115-mph gales with no damage. In February, 1943, the 24-in. main downwind bearing failed, and it took more than 24 wartime months before it could be replaced. The replacement was accomplished in March 1945. Some cracks were observed in the blade surfaces, probably due to stresses incurred by wind during the unit's two-year "locked" condition. Because of these cracks, it was decided to continue operation of the machine only long enough to finish a testing program then underway. It was intended that the unit then be shut down, to be fully repaired after the war. Unfortunately, one of the blades failed completely at 3:10 AM on March 26, 1945, due to propagation of those cracks. The 8-ton blade was thrown 750 ft from the tower. The project was deemed too expensive to continue.

Putnam had estimated that a block of six similar turbines could have been constructed on Lincoln Ridge in Vermont (a 4,000-ft elevation site) for about $190/kW of capacity. The economics of cheaper energy sources defeated the proposal, however. Thus ended the reign of the largest windmill ever constructed, a technical success that failed due to economic pressure and wartime production bans. Besides having generated an estimated 279,000 kWh of electric power during a single year,[67] the turbine project generated a large body of information on large-scale windmill operation, design theory, wind behavior characteristics and wind power conversion in general.

Although the interest of the Federal Power Commission (FPC) was sufficiently aroused for it to delegate Percy Thomas (the large-scale wind power proponent next in line to Putnam) to survey the possibilities, more than 10 years of work by the man (during which he stated that a 7.5-MW unit could be designed and built for $68/kW of capacity[67]) brought only promises. A hearing held in 1951 with the FPC and the Department of the Interior yielded an enthusiastic decision to design and construct Thomas' concept, but no real action was taken. Official government interest in wind turbines continued to be nonexistent until 1972.

THE NEW APPROACHES

The embryonic new philosophies concerning alternate life styles that sprang up during the late 1960s and early 1970s, progressive publications such as the *Whole Earth Catalog,* and the thousands of "no-gas" signs that appeared in 1973 during the oil embargo collectively revived interest (and hope) in the potential of wind energy conversion. Numerous new concepts have been built during this decade promising many significant advantages over the older designs of years ago. Further investigation has yet to test their potential.

Thomas E. Sweeney, director of the Advanced Flight Projects Laboratory at Princeton University, designed a new type of sail while a senior Princeton researcher. The "Sailwing" was later tried out on small planes through support by the Fairchild Corporation. Eventually, in 1972, Sweeney and others at Princeton built a Sailwing wind generator some 25 ft in diameter, utilizing two skeletal blades covered with Dacron polyester fabric (Figure 6.3). The horizontal axis machine was able to generate 7 kW in a 20-mph wind. The light and inexpensive nature of the device is complemented by its apparent durability (very little damage sustained during "rough winter storms and high winds at Princeton University").[73] The Energy Systems Group of Grumman Aerospace Corporation is currently involved in developing a 5-kW capacity version of the Sailwing, that will probably use three wings instead of two and "produce about 12,500 kilowatt-hours per year with suitable winds, such as on Long Island. This could easily serve an average residence if the output is leveled and converted to 60 Hz."[73] Another sailwing-type design has been generated by M.M. Sherman of the New Alchemy Institute in Woods Hole, Massachusetts. The design was developed for use in unindustrialized regions of the world, such as India.[74]

Windworks, an organization formed in Wisconsin in 1970 by Hans Meyer, an engineer, and sponsored by R. Buckminster Fuller, has been looking hard at developing cheap, reliable, elegant small-scale wind conversion systems. An interesting concept is that of forming the blades with expandable paper honeycomb. Supposedly, fairly complex blade shapes can be fabricated from this material with common power tools at a very low cost. The material is then expanded on the shaft and covered with fiberglass cloth, to be filled and sanded before completion.[72,74] The Windworks personnel have, since their founding, worked closely with many other wind researchers, and the group is fast becoming a central hub for wind-related information and knowhow.

An airplane mechanic from Ocala, Florida, Thomas O. Chalk, has come up with a wind turbine configuration that has generated considerable

Figure 6.3. Sweeney's Sailwing wind turbine, using flexible fabric airfoils, rather than a more rigid material. Wind forces cause the blade shape to change during operation.

enthusiasm as well as electricity. The structural strength, light weight and low cost of Chalk's turbine wheel is accomplished with a method of construction closely related to a bicycle wheel (Figure 6.4). The efficiency of this configuration is reportedly very high, and American Wind Turbine, Inc. was formed in Oklahoma in 1976 and currently markets systems based on this design.[72,73] It is felt that no insurmountable problems would be encountered in scaling this type of wheel up into the megawatt generation range.

An engineer from San Diego, Edmund L. Salter, has come up with yet another basic configuration. His idea is to situate three relatively large rotors (each having three blades) with peripheral rims tangentially around a smaller central-spoked wheel (Figure 6.5). As the three larger wheels revolve about their fixed centers, their rims supply kinetic energy directly to the rim of the small wheel. The small wheel's shaft then drives a generator. The prototype had a reported generation capacity of 7 kW in a 28-mph wind. Salter has formed a manufacturing company to mass-produce these devices and reportedly will be producing units significantly less costly than the only other machine available in the 630-kWh/month power range (the Swiss Elecktro).[72]

THE ERDA-NASA MODEL ZERO

The largest piece of equipment used at present to look at wind conversion is also the federal government's first major attempt in the field.

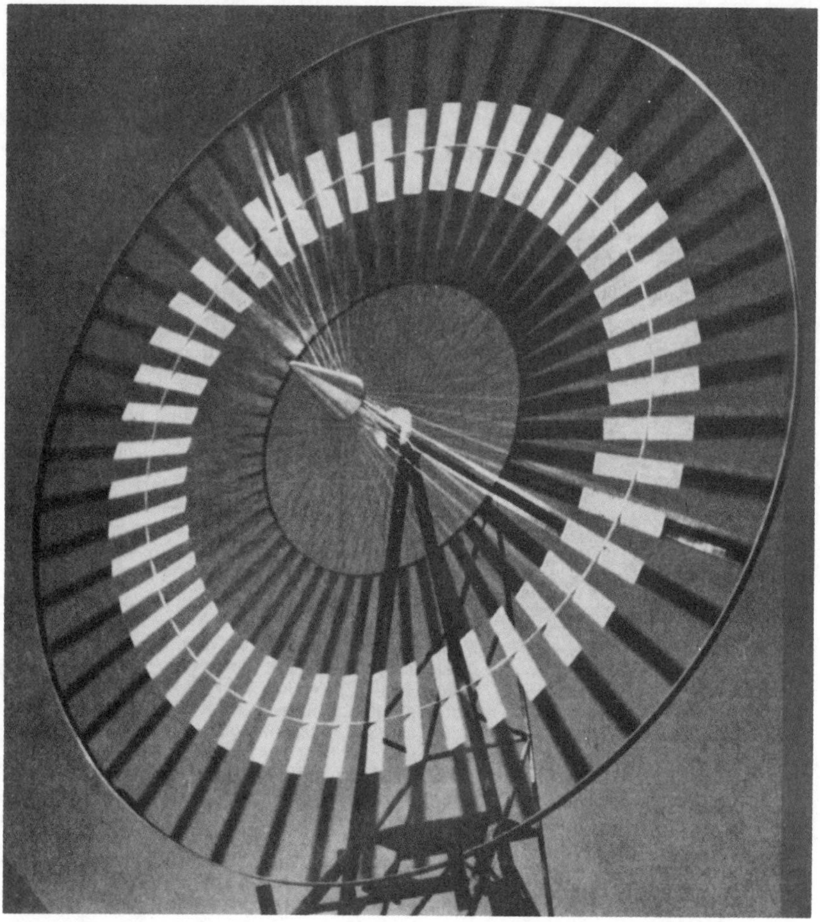

Figure 6.4. The turbine wheel developed by Chalk. Its configuration is unique in wind rotor applications, and shows a great deal of technical and commercial promise.

The machine has been installed at the Lewis Research Center's Plum Brook Station near Sandusky, Ohio. Construction was completed in September of 1975. As shown in Figure 6.6, the Model-Zero is a two-bladed machine mounted on an open truss tower. The tower is a 100-ft-tall, pinned-truss tower sunk into a concrete footing. This kind of tower was chosen for its contribution of low cost, easy maintenance access and its high natural resonance frequency to the project.

The blades are aluminum, each 62.5 ft in length, weigh 2,000 lb and are mounted downwind of the housing (in a fashion similar to the

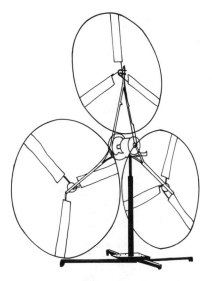

Figure 6.5. Another unique configuration, Salter's rotor has its own inherent advantages.

Smith-Putnam turbine and many European designs). This arrangement eliminates the possibility of any blade-tower collisions, while also allowing for a lesser machine cantilever at the top of the tower. The blades are tapered from the hub to their tips and have a total distributed twist of 26.5°,[75] allowing the tips (that will cut the air a great deal faster than the hub ends of the blades) to attack the wind at a much shallower angle.

The turbine will begin to rotate when wind speed reaches about 8 mph, and will reach its full rated 100-kW output at a wind speed of 18 mph. At this point, the losses in the mechanical transmission train "require that the rotor turbine extract 133 kWe from the 18 mph wind to ensure (rated)...output."[74] The maximum design rotation speed of the turbine is 40 rpm, a speed that will be maintained in wind speeds up to 40 mph by mechanically feathering the blades to "spill" excess power. When the wind speed reaches or exceeds 60 mph, the blades are fully feathered and so will not revolve. The rotor drives an 1,800-rpm alternator set up to deliver 60 cycles per second, three-phase power. This alternator is situated at the top of the tower to avoid long rotating shafts. The alternator's higher rmps are achieved with a step-up gear box.

All dynamic structures, especially on this kind of scale, are subject to wind-induced vibrations. If, during operations or shutdown, the vibrations occuring in the blades were of the same frequency as the vibrations in the tower structure, the cyclic forces would tend to reinforce each other, and

Figure 6.6. The ERDA/NASA Mod-O Wind Turbine. The first large wind machine ever built with taxpayer's money, this installation is only the first in a series of projects concerned with wind energy conversion. Two rotor blades spanning 125 ft were hoisted into place atop a 100-ft tower completing the 100-kW windmill located at Lewis' Plum Brook Station. The windmill project is a part of the Wind Energy Program at ERDA and will serve to study the potential of wind as an alternative source of energy. Once in operation, the blades will start turning in an 8-mph wind and reach maximum kilowatt output in a 19-mph wind. The experimental windmill is expected to generate enough electricity to power about 30 homes. (Courtesy NASA.)

could tear the structure apart. As more is learned about wind machines and their operation, it will become easier to predict the vibrational behaviors that will occur. For this project, it was decided to play it as safe as possible.

During a year of operation, estimates have it that the turbine will generate approximately 180,000 kWh electricity, about enough to supply power to 30 residences.[67] The project is designed to provide good hard data on wind conversion, to serve in the future as a test bed for other pieces of equipment, to help determine if this power source can be cost-effectively exploited, and will act as a foundation on which subsequent work on wind power utilization may be based. If the Model-O proves favorable, work will continue on towards construction of a Model-OA, a turbine of the same size, but capable of producing 200 kW, due to design differences in the mechanical and electronic subsystems. This project will give more experience on interfacing with utility networks. Three Mod-OA's are planned. A Mod-1 is scheduled to begin operation late in 1978, will be rated at 2,000 kW, and will have a rotor 200 ft in diameter (temporarily becoming the largest U.S. wind machine in history). The largest unit in the program, the Mod-2, is slated for operation commencement in late 1979, will be rated at 2500 kW and be designed to operate its 300-ft diameter (or larger) rotor in a site having a mean wind speed of 14 mph (this is a mean wind speed common to many sites throughout the United States).[58]

The Mod-O project cost about $985,000, and system cost per unit capacity has been estimated at about $6,500/kW.[67] This cost, as with all experiments that are developed for mass production, will go down substantially.

VERTICAL-AXIS MACHINES

Although the first definite records of wind conversion described Persian windmills having a vertical axis, with paddles that revolved like a merry-go-round, very little has been done to develop the vertical axis concept, compared to the manpower injected into the type described above. Some of the homemade machines used in America during the last century used vertical axes as well.

Technological principles were first deliberately applied to increase the efficiency of vertical axis machines by Commander S.J. Savonius of Finland. In 1929 Savonius patented a rotor, based on experiments that he had undertaken, that resembled (in cross section) an "S" shape. The basic configuration is shown in Figure 6.7 and was used in early years mainly in "gimmick"-type applications (such as unusual storefront signs), if at all.

The Savonius concept has, however, been used for many years to rotate ventilator hoods on the roofs of buildings.[76] Another device has been applied successfully as an ocean current metering device.[74]

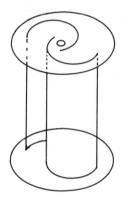

Figure 6.7. The Savonius rotor has many possible applications and is one of the simplest wind turbines to fabricate.

As with all vertical axis wind machines, the Savonius rotor instantaneously responds to winds coming from any direction, and changes in wind direction have no effect upon its performance. This gives the vertical axis wind machines their primary advantage: the absence of a need for mechanisms to orient them towards the wind. Figure 6.8 shows the airflow through one of these rotors. Regardless of the wind direction relative to the device, one of the semicylinders will present a concave surface to it. As the wind enters this rotor, some of its kinetic energy serves to drive the concave surface "away" (in a clockwise direction). The wind then continues on past the central axle while it changes direction, contributing more of its energy to drive the other half of the rotor in the "other direction" (which is also clockwise). As can be seen, the wind then exits through the rotor's open side. This sketch shows a wind tunnel situation, in which the rotor was fixed in one position. While this illustrates the rotor's driving principle, it should be remembered that the actual dynamic situation is much more complex. The original Savonius experiments exhibited "fairly high" rotor efficiencies of about 31%, "but was inefficient per unit of weight, since all the area swept was occupied by metal." [71] More recently, this type of rotor is being actively explored by Elecktro GMBH Company of Switzerland,[74] M. Hackleman and others of Earthmind in California,[77] and also by Sandia Laboratories in New Mexico.[67]

The other major type of vertical axis machine was patented within two years of the Savonius device, in 1931, by G.J.M. Darrieus of France. As Figure 6.9 illustrates, the original Darrieus rotor had two curved, flat "jump rope"-shaped blades (the more recent term for this shape is

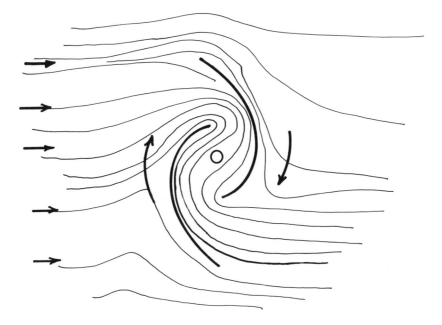

Figure 6.8. The airflow through a Savonius rotor clearly illustrates the principle behind its operation.

"troposkein," signifying the shape assumed by a cable when suspended from two points) that revolved around the central axis. This arrangement is also nondirectional, but involves much more complex aerodynamics than does the Savonius, resulting in additional problems. Typical versions thus far constructed of the Darrieus rotor will not begin to rotate until wind speed reaches about 12 mph, showing that the rotor requires a relatively high starting torque. As the wind speeds up after rotor movement begins, the rotor will speed up to about 13 rpm, a rotational speed at which it will remain, no matter how hard the wind blows. Paradoxically, if supplemental force is used to speed the rotor up to about 65 rpm, the device will speed up to over 200 rpm on its own.[78] A clear understanding of this behavior has not yet emerged, and will not, without further systematic experience with the device.

In the early 1970s, Raj Rangi and Peter South of Canada's National Research Council worked on an improved version of the Darrieus or "eggbeater" wind turbine. They succeeded in generating 900 W with a machine having a 15-ft diameter rotor (the peripheral speed of the outermost section of the blades was some six times as great as the wind speed).[72] Impressed by the design, Dr. John D. Buckley and his colleagues at NASA's

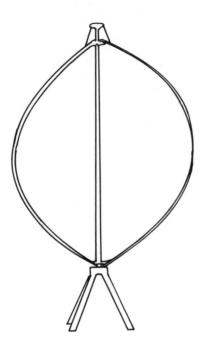

Figure 6.9. The simple shape of the Darrieus wind turbine belies the complex aerodynamics that make it work. Once those aerodynamics are understood,
it is expected that the capital costs involved with Darrieus turbines will
be quite low.

Langley Research Center built a similar wind turbine soon afterwards that
achieved similar performance.

The most sophisticated of the vertical axis wind turbines (VAWT) is
currently being tested at Sandia Laboratories, and is actually a composite
of the Savonius and Darrieus devices. It is of approximately the same
scale as the Canadian and Langley prototypes (Figure 6.10) but uses three
blades instead of two, and the blades are each composed of two straight
end sections and a middle troposkein section. The machine was built by
Ben Blackwell, Louis Feltz and Randall Maydew, of Sandia's Aerodynamics
Projects Department, who added a Savonius rotor to the top and bottom
ends of the central shaft to supply additional starting torque to the device.
Although the original intention of the group was to look at small wind
conversion applications, Blackwell has said that Sandia is now "hoping to
present a technological base to the ERDA (now the DOE) so that the
Darrieus turbine can be considered as an alternate concept for producing
large quantities of electricity to feed into the existing power grid. We

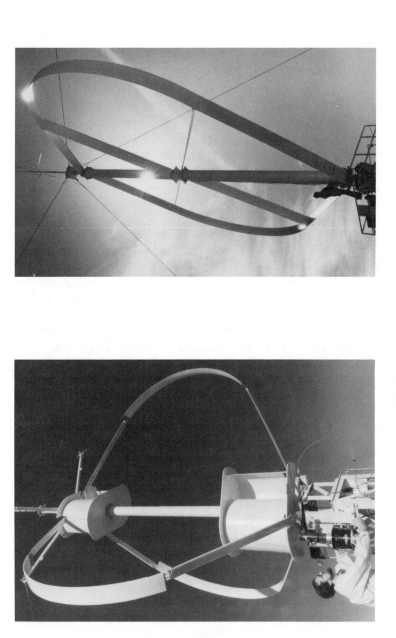

Figure 6.10. Sandia's VAWT is actually a hybrid of the Darrieus and Savonius concepts: **a)** Note the Savonius "buckets" placed at the top and bottom ends of the axle to help supply the device with starting torque. The reason for the "eggbeater" nickname is obvious; **b)** A much larger (50-ft diameter) Darrieus has recently been constructed by Sandia, and is now undergoing preliminary testing (Photos courtesy of Sandia Laboratories, Albuquerque, a not-for-profit prime contractor to the Department of Energy.)

think this turbine has the potential to do it more economically than the propeller system." [72]

OTHER IDEAS

A sizable number of other approaches have been theorized on wind conversion, but none of them have thus far gotten much beyond paper conceptualization. The Advanced Concepts Division of an organization in La Jolla, California, Science Applications, Inc., has proposed the construction of an artificial version of the solar energy/wind energy phenomenon to drive Savonius rotors. An enclosed volume of air would be subject to augmented solar heating, resulting in increased air convection. As the hot air rushes out of the enclosure, "cooler air would rush in through slots in the walls." [72] The Savonius rotors would be mounted in these slots. No further action on this concept has been reported.

West Virginia University has been involved in the development of both "obstruction-type" and "vortex-type" wind energy concentrators. The first type utilizes the fact that the air forced to flow around an obstruction speeds up considerably as it deflects. Wind tunnel testing has been undertaken on a rudimentary basis, and must be expanded on to allow sophisticated designs to be generated.

The latter type exploits a phenomenon exhibited in nature by tornadoes. As the air flows in a curved pattern, a pressure gradient is set up in a radial direction, resulting in a region of low pressure in the center of the vortex. Small vortices are set up when an aircraft wing slices through the air, and a device that utilizes this arrangement has been investigated by Sforza. [79,80,81]

Dr. James Yen, a fluid dynamics engineer at Grumman Aerospace Corporation, has taken the vortex concept a significant step further. His solution to the problem of wind's low-energy density is to let the wind drive a "tornado turbine." The concept is envisioned as a large unmoving tower with a comparatively small, vertically mounted propeller turbine situated at the bottom (Figure 6.11). The walls of the tower contain adjustable air vanes. As wind approaches the tower from any one direction, the vanes on that side of the tower only will be opened, causing the wind to enter the tower and to swirl around inside. As explained above, this situation causes a partial vacuum to form in the tower's center. The resulting suction will draw outside air in through the concentrating base of the tower, driving the fan. Unlike a more conventional windmill, in which only the air passing through the area swept by the rotor is involved, the effective collection area of the tornado turbine is slightly more than the entire frontal area of the tower. Also, since pressure differences rather than wind speed alone drive the turbine, it is theoretically possible to obtain more than 59.3% of the wind's power during operation. Yen has suggested that a power output of 1MW may be possible from a turbine only 6 or 7 ft in diameter (as opposed to a conventional rotor some 30 time larger). He has further estimated that capital costs for this form of wind plant could run around $500/kW or less. These projected costs, the stationary aspect of the outer structure, the conceptual simplicity of the device, its adaptability (solar energy could be used to induce a thermal updraft to drive the turbine) and its potential for aesthetic compatability (the tower could be made to look surprisingly like a building) indicate that this approach to wind conversion holds much promise. A proof-of-concept project is under development.[67,82]

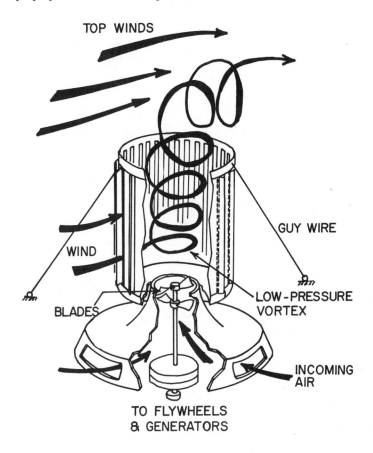

Figure 6.11. The tornado turbine concept brought forth by Yen. One of the more imaginative wind conversion approaches, it bears little resemblance to any other wind-gathering concept.

STORAGE AND CONVERSION FOR USE

The immediate accomplishment of all these machines is to convert the wind's kinetic energy into another form of mechanical energy, expressed as a rotating shaft. Since the principal application of wind energy will be to generate electricity, further changes in the output's form and availability are necessary. As stated by Wilkerson,[83] most alternate sources of energy (solar and wind energy included) are prone to the following common disadvantages:

1. Variability – general usage of energy requires a steady baseline capability. A reasonably constant power level is difficult, if not impossible, to provide with direct alternate energy source output.
2. Availability – energy consumption demands do not always coincide with the times at which energy is available from the alternate energy source.
3. Usability – mechanical (from wind), thermal (from solar thermal) or direct current (from solar cells) energy may not be used to directly power, say, a household electrical appliance. The energy must be converted to ac electrical power of the appropriate frequency and voltage.

Obviously, the wind machine itself can supply only mechanical power, of a variable nature, during periods of sufficient wind activity. To minimize the disadvantages listed above, more components will be necessary to comprise a complete workable wind energy conversion system. The main subsystems required include the mechanical wind machine itself (support structures, rotors, hub, shafts, gears, etc.); the generating system (to convert mechanical into electrical energy); the storage system (to retain excess energy for use during low-wind periods); the inversion system (to convert the variable dc from the generator to synchronous ac for consumption); and, of course, the backup energy source (the local power grid or onsite, fossil-fueled generator).

The devices discussed thus far in this chapter are the front-end (mechanical) systems used to gather the energy contained within the wind kinesis. The resultant shaft rotation is used to drive the subsequent conversion and storage subsystems. Initial conversion of mechanical to electrical energy will be accomplished by moving a conductor through a magnetic field (this is a functional description of a generator). The generator can and will be tailored for optimum effectiveness, depending on the operating conditions and characteristics of the rest of the system (wind speed, rotor type, demand parameters, etc.). Essentially, all of the necessary generation technology already exists.

Storage alternatives are many, with each at a different stage of development. The School of Electrical Engineering at Oklahoma State University has explored the theoretical possibilities of a number of storage methods, and has indicated that "high-pressure moderate temperature electrolysis" of water to produce hydrogen and oxygen holds the most promise.[74] Of the electrodes examined, a solid nickel finned type yielded the best results (electrolysis efficiencies of 85-90% were reported). Efficient electrolysis systems can be designed, fabricated and employed with current knowledge. The gases could be stored in underground caverns (with a low leakage rate of less than 1%/yr),[84] and later used to feed fuel cells or burner-turbine generators for purposes of electricity production.[74]

Compressed air could also be used as the storage media in those caverns, and later used to drive a turbine generator for electricity.[84] The wind machine could be used instead to directly drive a pump to provide hydro-electric storage (similar, in effect, to the Smith-Putnam turbine arrangement of over 30 years ago). Liquid (cryogenic) hydrogen could also be stored for eventual combustion. A superflywheel may be a suitable method for storing mechanical power, an approach being researched at John Hopkins University.[72]

Chemical storage alternatives are being looked at by a great number of researchers, and the concepts include batteries, oxidation-reduction cells and fuel cells. The many chemical battery types include lead-acid (similar chemically to automobile batteries), lead-cadmium, nickel-iron, nickel-zinc, nickel-hydrogen, nickel-air, zinc-air, zinc-chlorine, sodium-sulfur, lithium-sulfur and lithium-chlorine, some of which exhibit very high lifetimes and energy density capabilities.[72,74] NASA has been exploring the feasibility of electrically rechargeable "redox" flow cells, which chemically store energy with theoretically less cost and complexity.[72] Fuel cells could then produce electricity very efficiently when supplied with previously stored hydrogen and oxygen. Inversion equipment would be used to make the final conversion from dc to the appropriate ac energy, completing the process.

One more major concept could effectively satisfy all requirements, however. The idea is to feed synchronously inverted excess power directly into the nearby power distribution grid. The synchronous inverter is a device whose function is to convert variable dc power into ac power "at a rate synchronized to the frequency of the power grid."[84] From the point of view of the wind machine operator, the power grid behaves like a very efficient, infinitely large storage system, absorbing all excess output power and trading it back when necessary with very little energy loss. From the utility's point of view, the windmill's input energy is instantaneously distributed to utility customers, allowing the utility to consume that much less central plant fuel to satisfy consumer demand. Of the many options now being developed for wind energy conversion, "synchronous inversion provides the simplest and least costly approach to interfacing intermittent sources while still maintaining the high standards of power quality that utilities meet."[85] Of course, you must have ready access to the local power grid, as well as the benefit of a cooperative utility company.

It can be seen that, of the dozens of subsystems and approaches currently proposed, literally uncountable wind energy conversion system combinations are possible. Research performed to date has gone far towards proving wind energy's feasibility, and has indicated some general directions that look most promising. A great deal more research and development will be required before widespread, optimum wind energy conversion systems are a reality. Recent estimates are that some 1.5 million kilowatt of wind-derived electricity could be generated annually in the U.S. by the year 2,000, if wind systems were located at strategic locations, amounting to some 2.7% of the United States' projected total energy use in that year.[70,86] Wind, then, will not be our main source of energy, but would by all means be worth whatever research funding will be necessary to achieve it. Putting it another way, it will be less painful to allocate the funding now than to be without wind's potential energy contribution a generation from now.

For further reading and reference the reader is referred to "Fundamentals of Wind Energy" by Dr. Nicholas Cheremisinoff (Ann Arbor, MI: Ann Arbor Science Publishers, Inc., 1977).

CHAPTER 7

OCEAN THERMAL GRADIENT POWER

INTRODUCTION

More than 70% of the surface of this planet is covered with oceans. Of the total solar energy that reaches the earth's surface, approximately 45% is absorbed by the surface layers of these oceans, especially those in the tropical regions (due to the higher angle of incidence).[87] While some of this impounded heat is later given off to the atmosphere to drive the weather, the majority of it perpetrates a considerable temperature difference between the warm oceanic surface layer which extends to a depth of 150-500 ft and the colder water masses some 3,000 ft or more below. Due to the density gradient between the surface layer and the deep water, very little mixing occurs between the layers. Of course, this temperature differential is greatest in the equatorial regions of earth, so there is a constant reservoir of chilled water lying under the warm tropical and subtropical oceans and seas.[67] This circumstance has existed for eons, and, in fact, is a natural situation basic to our world, and comprises another case of solar energy once removed.

The main thermodynamic requirement for the operation of any heat engine cycle is the presence of a heat source and a heat sink. It is these two regions of "hot" and "cold" that put the cycle's working fluid through its phase-changing paces, allowing the engine to generate power. In most conventional power cycles, the heat source has been provided through the combustion of some kind of fossil fuel, giving rise to a large difference in temperature between the source and sink. While this large ΔT is usually advantageous (yielding higher efficiencies of operation), it is certainly not technically essential to heat engine operation.

Between the surface layers and deep (1,500-4,000 ft) subsurface layers of the oceans, within 20° north or south of the equator, annually averaged temperature of about 32-43°F can be expected.[88] The formal concept of exploiting this natural ΔT for purposes of performing work was first brought forth in 1881 by D'Arsonval, Campbell, Dornig and Boggia, from France, America and Italy, respectively.[89] The idea was that of a closed-cycle turbine run by an evaporated (working) fluid. The ocean's surface water would "heat and cause evaporation of a "working fluid"...which would pass through a turbine, thereafter being condensed by cold water pumped from deep layers and again fed into the evaporator."[91] This is the basic concept being most heavily investigated at present (that of a closed-cycle system), and is illustrated in Figure 7.1. As can be inferred from this discussion, ocean thermal energy conversion (OTEC) is fundamentally the conversion of the sea's thermal energy to a fluid's kinetic energy (phase changes with resultant volume and pressure kinesis), which, in turn, produces a turbine-generated mechanical output, and allows a generator to produce electrical power. This thermal-to-mechanical-to-electrical conversion is common to all OTEC systems, but can be accomplished in a number of ways.

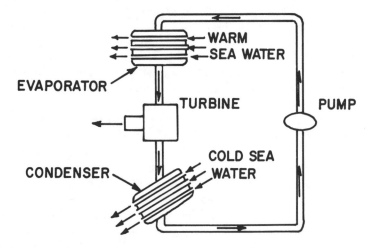

Figure 7.1. A very basic diagram of the operation of a closed-cycle OTEC system. The working fluid is most often ammonia or propane.

Another major type of ocean thermal energy conversion system is the open-cycle system, in which the seawater itself is used as the working fluid (a region of partial vacuum allows the warm water to vaporize). This approach was used in the first hardware work performed on OTEC development. A pupil of D'Arsonval, French engineer Georges Claude built a plant at Ougree in Belgium in the late 1920s that generated power by "flash-evaporating" warm seawater in a region of low pressure. The low pressure steam was used to drive a 3-ft-diameter turbine, generating 60 kW at 5,000 rpm, utilizing a total ΔT of about 35°F.[90]

This equipment was later transported to Cuba, erected on a similar shore location, and supplied with cold water by means of a 5.5-ft diameter pipe some 1.25 miles long. The pipe extended to a depth of about 2,300 ft, and provided a ΔT of about 25°F. The turbine used in Cuba produced 22 kW, but had too small a capacity in comparison to the other plant components. In fact, the pumping power requirements of this missized system exceeded 22 kW, causing the system to fail in achieving break-even operation. Claude, operating predominantly with his own funds, could not afford to purchase a more suitable turbine, having suffered heavy losses when the first two attempts at deploying a cold water inlet pipe had failed. Nonetheless, the technical feasibility of OTEC had been demonstrated successfully, paving the way for subsequent work.[88,90,91]

Some research of a less comprehensive nature was performed in 1931 by a French maritime company exploring the technique as a means of producing freshwater (the evaporation of seawater in an open cycle OTEC system is tantamount to distillation). In 1941 French government-funded research was undertaken to explore environmental impact and corrosion/biofouling problems (biofouling is the buildup of a biological film on the heat transfer surfaces of the evaporator and condenser – greatly curtailing OTEC efficiency). The University of California began work in 1951 on water desalination OTEC systems, estimating that their open-cycle design could "desalt water at a lower cost than any then or later developed system."[90]

However, the first major attempt at progress towards commercial OTEC started in 1960. The father and son team of Hilbert and James Anderson developed a closed-cycle concept incorporating a floating platform* to support water intakes and the heat exchangers, which are submerged to depths that provide external (seawater) pressure near those of the internal pressures of the working fluid.[91] A Freon refrigerant was eventually selected for this purpose. Adjusted to 1975 dollars, the Andersons estimated in the mid-1960s, total capital costs of the 100-MW plant would be about $330/kW of capacity, and that the busbar (ready for transmission to shore) power cost would be 6 mill/kWh (a mill is one-tenth of a cent).[91]

The National Science Foundation picked up on the Anderson's initiative in the early 1970s and contracted the University of Massachusetts and Carnegie-Mellon University to research OTEC possibilities. The concept put forth by Heronemus and McGowan of the University of Massachusetts

---

*Unlike Claude's approach, all modern conceptual OTEC systems are designed to float offshore, independent of land for the support of any equipment.

describes a 400-MW plant of submerged twin catamaran configuration. They have designed it to be operated off of the lower eastern coast of the U.S., in the Gulf Stream, with a ΔT of 32°F. Plate-and-fin-type heat exchangers constructed of 90/10 copper-nickel alloy are used in both the evaporator and condenser modules. Propane was chosen to be the working fluid in this closed-cycle system, and the elliptically cross-sectioned cold water feed pipe would be made of aluminum and extend to a depth of some 2,000 ft.[91,92] The Gulf Stream site was chosen for its relative ease of plant-to-shore power transmission by the use of submarine cables. In 1974 they estimated that capital costs would run about \$800/kW capacity, with a power cost (after the energy reaches shore) of some 15 mill/kWh. This concept, as are all the others, is continually undergoing revisions and iterations, with many alternatives being considered. Four different University of Massachusetts OTEC configurations have been developed on paper thus far.

Lavi and Zener of Carnegie-Mellon University allocated most of their effort toward the design of specific optimized components, especially the heat exchangers. In most designs, the huge required heat exchangers are "the predominant cost factor (in OTEC plants), taking up approximately half the plant's capital cost."[92] The team advocates a fluted tube aluminum heat exchanger design which is expected to have considerably enhanced heat transfer performance. A 30-kW model will be tested in 1978.

The encouraging results of these studies prompted the NSF (and later ERDA) to solicit industrial participation in OTEC development in 1974.[88] The two main research teams that resulted from the solicitation included Lockheed Missiles and Space Company, and TRW Systems and Energy Group. Each team generated its own particular OTEC configuration. Both concepts may have the capability to produce, onboard, various energy-intensive materials.

The OTEC system conceived by Lockheed, in conjunction with Bechtel Corporation and T.Y. Lin Associates, is shown in Figure 7.2. The power plant consists of a central submarine platform to which four independently operating power modules are mounted. The cold water pipe is fabricated of reinforced concrete and telescopes down some 2,000 ft below the bottom of the platform, already approximately 590 ft below the water surface. The outside diameter of the pipe varies from 129 ft to 105 ft and, when fully extended, serves to channel over 56,000 cubic feet per second (ft^3/sec) of cool, deep seawater to the lower level of the main platform during operation.

Fundamentally, the power cycle functions as follows. Warm surface water is pumped through the evaporator in the top of each power module (Figure 7.2 b). Each evaporator is in essence an enormous heat exchanger

(72 ft in diameter). The warm water transfers immense quantities of its heat to liquid ammonia (Lockheed's working fluid) every second, causing that ammonia to vaporize. As the vapor expands, it is channeled past the wheels of two generator turbines, resulting in mechanical shaft output. The vapor is then condensed back into liquid form by passing through the cold-water-fed condenser at the bottom of the power module. The now-liquid ammonia is pumped back up to the evaporator, completing the cycle. If you've read Chapter 4, you should be reasonably familiar with this kind of operation, since it is quite similar to the absorptive cooling cycle used in buildings. In the OTEC application, however, the ammonia is used to provide kinetic energy for electric generation, rather than to simply absorb and reject heat for space conditioning. The motivating principles are similar, but the difference in scale is gigantic.

The Lockheed System is designed to be moored in a stationary position within power cable reach of shore, and to produce a net of almost 265 MW of baseline electricity on a continuous basis. Importantly, this baseline capability is one of the most attractive of OTEC's charms. The oceans, acting as an infinite heat source and sink, are unaffected by the atmospheric and diurnal variations that plague the other solar energy sources, and are able to supply its energy at least as reliably as any nuclear or fossil-fueled power plant. This advantage is, of course, inherent to all of the OTEC systems discussed here.

The capital cost of Lockheed's plant has been estimated at some 1,700/kW in 1976 dollars nearly 50% of which is allocated for the titanium heat exchangers), yielding a busbar power cost of 33 mill/kWh. This full-scale plant could, according to Lockheed, be in operation by the end of the next decade.[88,90-93]

A second concept has been generated by the team of TRW, Southwestern Engineering Co., the C.F. Braun Company, the Linde Division of Union Carbide Corporation, Carnegie-Mellon University and Global Marine Development, Inc. (Figure 7.3). Their OTEC system also uses ammonia as the working fluid, in a thermodynamic cycle similar to the Lockheed device (Figure 7.4). The cycle and its equipment are designed to operate with an available ΔT of about $40°F$, and several sites off the Gulf Coast, Puerto Rico, the Virgin Islands and the Hawaiian Islands have been found to provide this thermal requirement.

TRW's plant has a design net power output of 100 MW, generated by four power modules, this time positioned within the cylindrical hull's interior. The hull itself is slightly less than 340 ft in diameter and is fabricated of reinforced concrete, a material commonly chosen for this sort of application both for its relatively low expense and its long effective lifetime in a sea environment. The horizontal heat exchangers used

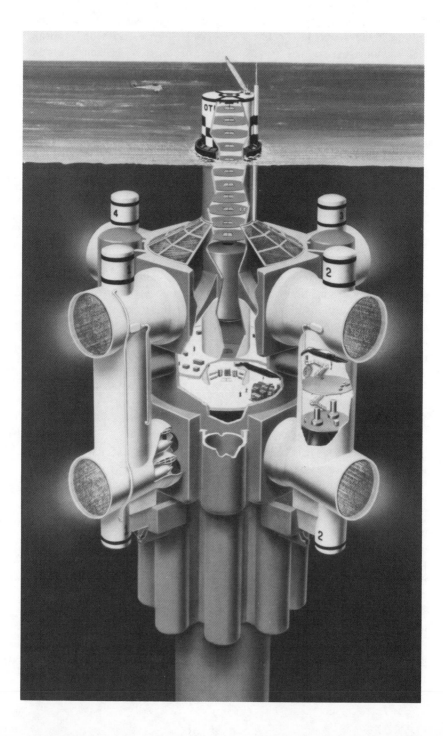

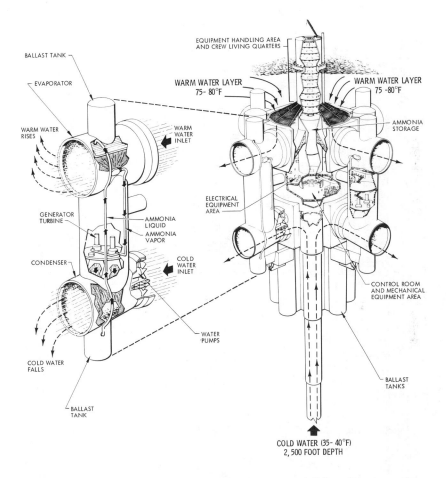

Figure 7.2. The Lockheed OTEC design **(a)** shows an artist's cutaway conception of the plant; and **(b)** Gives an idea of the functions of its component parts. Oceans are constantly collecting massive amounts of solar radiation and storing it as heat energy. A gigantic heat engine — such as diagrammed here — uses the warm surface water as the heat source, and the cold water from the depths as a heat sink. The warm water vaporizes a liquid, such as ammonia. Like steam, this gaseous, pressurized ammonia drives turbine-generators. The ammonia then is condensed to its liquid form by the cold ocean water, and this closed cycle continues. When fully developed, an OTEC heat engine is expected to supply significant amounts of electric power at competitive prices. (Courtesy Lockheed Missiles & Space Co., Inc.)

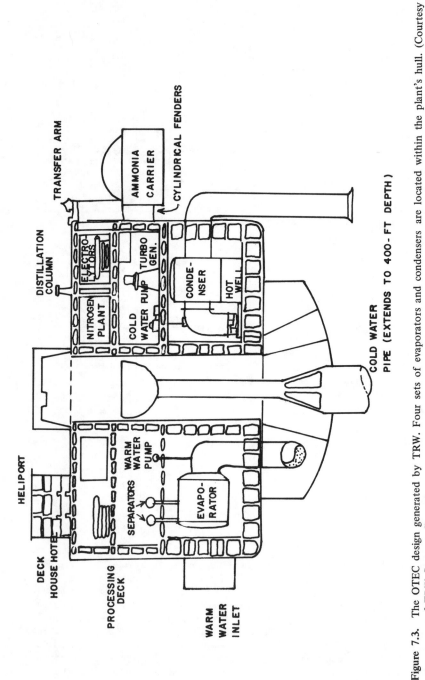

Figure 7.3. The OTEC design generated by TRW. Four sets of evaporators and condensers are located within the plant's hull. (Courtesy of TRW Systems Group.)

OTEC POWER SYSTEM

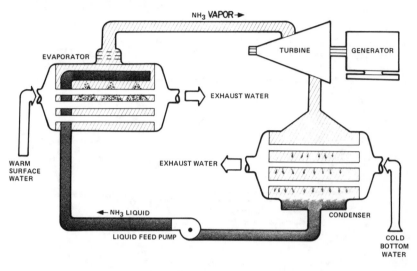

Figure 7.4. This diagram serves to give the reader more of a "feel" for how the closed OTEC cycle operates. (Courtesy of TRW Systems Group.)

in the condensers and evaporators use titanium tubes to circulate the sea-water (Figure 7.5).

The 50-ft-diameter cold water inlet pipe is made of fiber-reinforced plastic and reaches a depth of 4,000 ft. As with the other OTEC designs, TRW's plant will be constructed to handle the storms and currents encountered at sea. The position of the plant could very possibly be held by a deep-sea anchoring device, but the plant would also be capable of dynamically positioning itself by using the exiting pumped seawater as controllable thrust (this option may be more dependable and inexpensive). Study of the baseline TRW design yielded a capital cost estimate of $2,100 per kW in early 1975 dollars, of which almost half goes towards heat exchangers. The busbar cost estimate resultant from this is 35 mill/kWh.[91,94] Important here is the fact that, as with Lockheed and the others, present cost estimates are based on first or second "paper" versions of concepts, indicating that future power costs may very probably be significantly reduced by reiteration of power cycle operation, materials used, design improvement, and construction and operating experience. The value of comparing these reported cost estimates (which are also based upon various ΔTs and power transmission schemes) lies only in general ranges, rather than specific figures.

Another major contribution to the blooming OTEC embryo was made by the Applied Physics Laboratory of Johns Hopkins University, in a joint

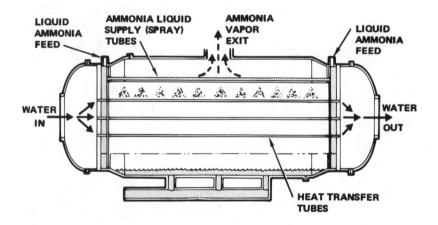

Figure 7.5. Shown above is a diagrammatic cross section of a horizontal tube evaporator. A generally similar type of component is used to condense the vapor back into liquid form. (Courtesy of TRW Systems Group.)

effort with the Sun Shipbuilding and Dry Dock Company, the Woods Hole Oceanographic Institution, Avondale Shipyards, and Kaiser Aluminum and Chemical Corporation, among others. The research was sponsored by the Maritime Administration (MarAd). Unlike the TRW or Lockheed configurations, the APL/JHU plant design has fundamentally a rectangular 200 x 475-ft hull, situated symmetrically above the cold water inlet pipe (Figure 7.6). The hull, again, is fabricated with reinforced conrete, and is designed for surface deployment. The evaporator and condenser heat exchangers have been designed for gravity flow past aluminum tubes, rather than for pump-induced forced convection. The component walls will serve doubly as heat exchanger shells and as ship structural members.

Cold water is pulled up through the 60-ft-diameter reinforced concrete pipe, from a depth of over 2,500 ft, to "head ponds" situated slightly above sea level. A similar feed approach is used with the warm surface water, and both flows are gravity discharged some 70 ft below the surface. The heat from the surface water drives the closed ammonia loop in a general fashion now familiar to you, the main difference being that the ammonia in this power cycle is not fully vaporized during the heating process. It is felt that the less than maximum power cycle efficiency is more than offset by the relative simplicity, capital savings and ease of maintenance afforded by the use of large diameter heat exchange tubes. This approach reportedly results in the lowest possible overall plant costs.

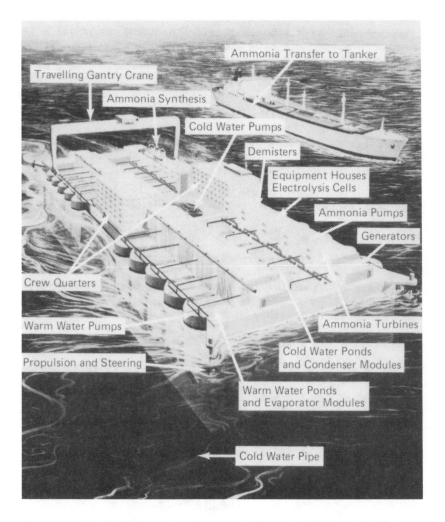

Figure 7.6. The APL/JHU concept for a "tropical grazing OTEC plant-ship" to produce liquid ammonia at sea.

This contention would seem to be well founded, since the capital cost estimate of the 100-MW plant indicates $1,148/kW (projected to decrease to $575/kW after several plants of 325-500 MW size have been modularly constructed). One additional significant factor has been cited as a contributory factor in the low cost, the capability of the APL/JHU OTEC facility to "graze." Research to date has indicated that seasonal variations in the available ΔT as well as plant exposure to high winds and waves could be minimized were the plant to follow some pattern of movement through an appropriate region of the Atlantic, optimizing plant performance.[91] It should be noted here that, whereas the Lockheed and TRW designs are essentially concerned with the application of present day technology, "the APL/JHU design concept is based on technological innovations in heat exchangers, hull and ammonia plant construction."[92] This ammonia plant will be discussed further on.

A boldly innovative concept for OTEC is being developed by Clarence Zener and John Fetkovich of Carnegie-Mellon University, and is called the "foam OTEC system" (Figure 7.7). Whereas all other OTEC systems use seawater as heat source and heat sink and discard that water after it gives up a few BTU/lb of thermal energy, the foam OTEC technique uses the water as the working medium (in a way different from Claude's version of the open cycle plant). Figure 7.8 is a sketch of a single foam cell. The physical barrier imposed by the liquid cell wall of each bubble serves to fully contain a unit quantity of water vapor, forcing the vapor and the bubble to move together. The foam reportedly liberates, during travel from a higher pressure region (0.5 psi) to one of lower pressure (0.1 psi), approximately 1.2 Btu of work per pound of foam.

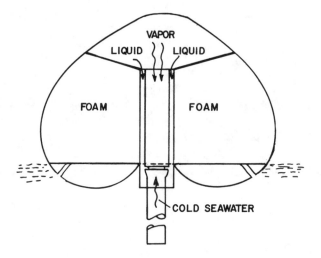

Figure 7.7. While the CMU design for a foam OTEC system is still in a rough conceptual stage, this sketch serves to illustrate its basic configuration.

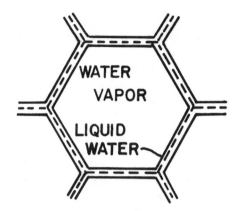

Figure 7.8. This represents the foam cell model used by Zener and Fetkovich during development of their concept. As shown, the water vapor contained within each cell is forced to move along with its liquid "envelope."

The reinforced concrete dome of this floating OTEC plant is designed so that the lower pressure region exists at the top of the dome, at which the open end of a 960-ft-high standing pipe is situated (this is the estimated maximum theoretical height). The lower pressure (caused by the low vapor pressure of the cold seawater) will cause all foam generated at sea level to rise to the pipe's open end. At this point, the foam will be "broken," releasing the foam's liquid and condensed vapor (through contact with a cold water spray) to flow down the pipe. This arrangement thereby creates an enclosed continuous "waterfall," with a head of 960 ft, to feed conventional hydraulic turbine generators at the bottom of the pipe.

Zener and Fetkovich have projected the dome to be 1,600 ft in diameter, and supported by a semidoughnut-shaped barge integral to the plant structure. The plant has an estimated generation capacity of 100 MW. Total capital costs have yet to be generated, but the simplicity of the system should make it a prime candidate for more detailed study.[95]

OTEC POWER APPLICATIONS

While the submarine cable transmission of electricity to the shore is the most direct application of OTEC output, it is by no means the only one. Electric power generation will, however, be discussed first.

ELECTRICITY BY CABLE TRANSMISSION

Studies indicate that OTEC power would immediately, on deployment in the late 1980s, "be close to competitive with coal-fired base load (utility) systems in the Southeastern United States."[96] The power would also be roughly competitive with nuclear fission-derived electricity.[91] The absence of any costs for fuel eliminates any future fuel price escalation problems, causing OTEC power to get more advantageous as time passes. In addition, OTEC is inherently a very safe methodology, with a maximum plant temperature of 80°F or so, and avoids any major risk to the environment through chemical or thermal pollution.[97] The concept also has a considerable potential for important breakthroughs and improvements, being an embryonic technology at present.

The main limitation to submarine cable transmission, then, is the requirement for the plant to operate in a fixed position reasonably near to shore. According to work done by Lockheed, the cost for transmitting power through a cable along a plant-to-shore distance of 100 miles is about 5 mill/kWh. If the cable must be longer than 200 miles, costs quickly rise beyond cost-effectiveness. Although numerous sites within this "leashed" region are available, the restriction of OTEC to only this type of usage would severely cut down its potential for productivity.

Another prime application of OTEC power concerns the production of certain energy-intensive materials on board the plant itself, with the sea-derived power being used onsite to drive manufacturing operations. The TRW and APL/JHU concepts take this function into consideration. In fact, the APL design is more a seagoing factory than solely a power plant. This promises to be a perfect mechanism with which to directly involve the manufacturing industries with the support of OTEC development.

ALUMINUM PRODUCTION

A specific use along these lines could be the reduction of alumina (made from bauxite on land) into aluminum. Estimates have it that the additional electrical capacity necessary to produce aluminum in this country by the year 2010 amounts to more than 100 million kW.[96] Alumina reduction would probably be the next OTEC application to become economically feasible, after the submarine cable alternative, and could supply the Atlantic, Pacific and Gulf Coast American markets with aluminum produced at sea. A portion of the aluminum could also be allocated for use in the construction of heat exchangers for subsequent OTEC plants. APL estimates that one tropically situated plant could produce enough aluminum in one year to satisfy the needs of 10-16 similarly sized units. Aluminum output could also be applied to uses such as widespread fabrication of flat-plate collectors, or to satisfy the automotive industry's expected increases in aluminum demand.

AMMONIA PRODUCTION

About 2.5% of the total supply of natural gas in this country is used to produce ammonia for use in fertilizers and other products. During the manufacturing process, approximately 34,000 standard cubic feet of natural gas are used to produce one ton of ammonia (NH_3).[91]

A simplified diagram of APL/JHU's OTEC ammonia synthesis process is shown in Figure 7.9. As reported by Dugger *et al.*,[91] this plant will require three main inputs: (1) thermal energy from seawater for electricity, (2) hydrogen from seawater, and (3) nitrogen from the atmosphere. Basically, electrical output from the OTEC power cycle is used on board to electrolize (electrically split) seawater into hydrogen (H_2) and oxygen (O_2). A simultaneous but separate process will serve to liquify ambient air and to extract nitrogen from it. "Downstream" of these processes, the hydrogen and nitrogen are combined, forming the ammonia end products. This material would be stored on board and periodically transferred to transport ships when necessary. If such a system were installed on the first full size, 325-MW APL/JHU plant, it is expected to add about $60 million to the total capital cost. NH_3 production costs would amount to some $130 per short tons of ammonia delivered to a U.S. port dropping to $100 per ton by the eighth or ninth plant, in 1975 dollars. An alternate use of this ammonia back in the U.S. could be to subsequently split it back to hydrogen and nitrogen, and to feed these gases, along with oxygen, into a fuel cell process.[91] In this way, the ammonia could be used as a "hydrogen carrier" for land-based electrical generation, and thus could allow OTEC to become a major factor in a hydrogen economy. Lockheed and TRW have also shown some interest in at-sea OTEC materials production. The large horizontal deck area of the TRW design would lend itself to such arrangements.

While not as economically attractive as the other materials, it would be technically possible to synthesize or process a number of other materials on board an OTEC plant. A study conducted by DSS Engineers, Inc. has resulted in a list of "high-demand, high-growth rate products,"[91] that could be feasibly marketed from OTEC productions. The materials include magnesium (present in seawater), ethylene oxide, polyethylene, chlorine, steel and potash, among others.[98] While costs of substantial accuracy are impossible to generate at present, it seems that production of materials other than aluminum and ammonia would be uncompetitive until cost reductions could be achieved, since the economics and some of the technical problems involved are "far more complex than the construction of the OTEC system itself."[97]

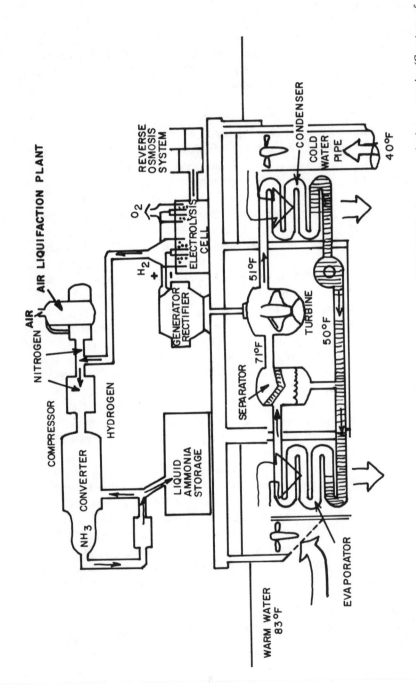

Figure 7.9. A schematic diagram of the systems that may be used on an APL/JHU OTEC plant to synthesize ammonia. (Courtesy of the Applied Physics Laboratory of Johns Hopkins University.)

MARICULTURE EFFECTS

A potentially large beneficial side effect of OTEC plant operation has not yet been discussed—that of mariculture. Off the northwestern coast of South America, a natural upwelling of the dissolved nutrients (nitrogen and phosphates) that normally lie inactive within the cooler oceanic layers, causes the existence of an extremely rich fishery region. Those nutrients, brought to the surface by the Humbolt current, "combine their fertility with the photosynthesis capability of the tropical sun," forming a solid base for a mariculture food chain, resulting in large catches of fish.[87]

This natural upwelling will be simulated on a somewhat smaller scale by every OTEC system that is deployed, suggesting a substantial potential for the indirect production of protein, in addition to any primary end products that the plant may output. Indeed, various land-based and floating open-cycle OTEC plants have been proposed that purposely provide for desalination and mariculture facilities in addition to power production.[92] Early studies have indicated that dramatic increases in fish populations would result throughout a 900-ft radius area surrounding a "small" 20-MW plant as a result of artificial upwellings caused by plant operation.[91] The main requirements for optimizing the mariculture enhancement potential of an OTEC plant would be to locate the plant in an area of small oceanic current, so that the dissipation of nutrients would be as small as possible.

In general, OTEC is the solar energy methodology that holds the most promise for large-scale reliable and implementable impact on the energy crunch in this country and around the equatorial regions of the world. Technically and economically, OTEC's status of development qualifies it as a prime contender for any future courses of action for baseline power generation and the production of certain materials. Expect to hear much more about it during the fairly near future. The Department of Energy expects to have at sea by late 1981 or early 1982, one or two pilot plants, producing about 10-MW each, to be followed by commercial plants as soon as successful operation has been demonstrated and industry begins to take a substantial share of the costs, eventually to take over completely.[99]

CHAPTER 8

CHEMICAL CONVERSION OF SOLAR ENERGY

INTRODUCTION

Aside from its naturally accomplished synthesis, the great technological advantage of a fossil fuel lies in its form, that of a kinetically inert chemical compound. Power is contained within these compounds as chemical potential energy, and is locked into the very structure of the material. This form allows the fuel to be concentrated and stored indefinitely, requiring some activation energy to be applied for the release of this potential (the activation is normally achieved with a spark or flame to initiate thermal combustion). This situation avoids the problems with leakage and quality degradation that are always associated with thermal or electrical forms of energy storage. In other words, chemical potential energy, properly handled, can be stored in a way to eliminate the possibility of a premature release of its power. This is the goal of energy storage: what you put in is waiting for you to use later on.

Previous chapters have discussed the conversion of solar energy, by various methods, into thermal, mechanical and electrical forms. Only in Chapter 7 was a solar conversion system described which was designed to produce a chemical end product (hydrogen or ammonia). This end product was achieved, moreover, by way of a circuitous, multistepped, solar-to-thermal-to-mechanical-to-electrical-to-chemical process.

The most direct method discussed thus far of converting solar into electrical energy is by means of the solid-state photovoltaic devices that filled Chapter 5. It was shown that, while the solar cells have demonstrated reasonable efficiencies, their high sensitivity to impurities and the resultant high costs of fabrication have placed them in a presently noncompetitive economic position. Large strides in cell manufacturing economy and in the lengthening of operating lifetimes will be necessary prior to the widespread use of these devices as a source of power.

In the last decade, a new group of experimental devices has been proposed that promises to allow for a direct and more economical method with which to accomplish conversion of solar energy into electricity and/ or chemical potential energy. They fall into the category "photoelectrochemical" devices, and are of two subtypes. One directly converts solar energy into fuel (chemical potential energy) and the other can direct solar-to-electrical conversion by a method less costly than the previously described solar cells. These two groups of devices are called photoelectrolysis cells and semiconductor-electrolyte cells, respectively. It is discussions of these devices and the general principles behind them that comprise this chapter.

PHOTOELECTROLYSIS

In 1969 two Japanese researchers, Fujishima and Honda, began to publish the findings resultant from their work with a certain type of electrolysis cell. Their reports soon caused the emergence of a full new branch of alternate energy research. The two men had connected a wafer of n-type titanium dioxide (TiO_2) to a platinum counterelectrode, both of which were immersed in an aqueous electrochemical cell (Figure 8.1). They then exposed the TiO_2 electrode to an emission of ultraviolet light, thus causing a photovoltage (similar to that described in Chapter 5) to occur in the cell sufficient to dissociate water into its constituent hydrogen and oxygen.[100] The main significance of this finding and their subsequent work was the resulting possibility that a similar, more developed, cell may be able to split water with solar insolation as the sole energy input. The achievement of such a device would allow for large-scale, inexpensive, reliable production of hydrogen.

To become somewhat more conversant with this concept, some further discussion is appropriate. Electrolysis is a process in which electric current is passed between two electrodes and through an electrolyte, causing a chemical change within that electrolyte (the electrolyte is a nonmetallic solid or liquid that can support the flow of ions in an electric circuit). Figure 8.2 is a basic representation of an electrolytic cell that would be used to electrolize water (the electrolyte in this case) into H_2 and O_2. In simple electrolysis, power for the electric current is supplied by an external source. During operation, the induced chemical reactions occur at the interface between each electrode and the electrolyte. At the anode (the electrode in the right in Figure 8.2), electrons are removed from the electrolyte to flow through the external circuit. This removal of electrons is known as "oxidation" and results in the evolution of oxygen gas from the chemically charged water in the immediate vicinity of the anode. At

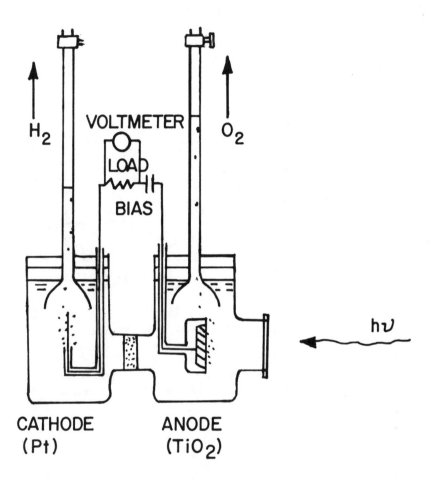

Figure 8.1. This, in schematic form, is the classical photoelectrolysis cell used in most research on the involved principles. One of the goals of this research is to eliminate the need for the externally applied bias potential indicated. The electrode materials shown are the ones used by Fujishima and Honda.

the cathode (the electrode on the left), electrons are injected into the electrolyte, thereby "reducing" it. This reduction causes hydrogen gas to be generated at the cathode.

Another way to think of this process is to imagine water as an equilibrium mixture of negative hydroxyl (OH^-) and positive hydrogen (H^+) ions, so that:

$$H_2O \overset{\rightarrow}{\leftarrow} OH^- + H^+$$

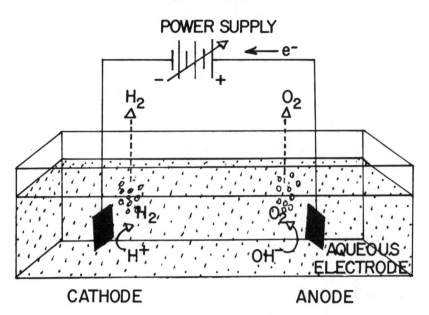

POWER SUPPLY

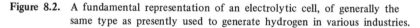

CATHODE　　　　　　　　　**ANODE**

Figure 8.2. A fundamental representation of an electrolytic cell, of generally the
same type as presently used to generate hydrogen in various industries.

In this way, it is easier to understand that the oxidation-reduction
(redox) reactions that occur at the electrodes are primarily affecting components of the water electrolyte. This situation can be represented, as
discussed by Wrighton,[101] as follows:

$$4H^+ + 4e^- \rightarrow 2H_2 \qquad \text{(cathode process)}$$
$$4OH^- \qquad \rightarrow O_2 + 2H_2O + 4e^- \text{ (anode process)}$$

In the cell, the external load applies a difference in voltage (potential)
across the electrodes, which is necessary to cause electrons to flow properly in the aqueous electrolyte to allow electrolysis. The theoretically
required potential difference is 1.23 V, with water as the electrolyte, but
it has been found that about 1.5 V is necessary for continuous satisfactory
operation. The last formula listed above is a standard representation of
the electrolysis of water. It indicates the method currently used to produce
hydrogen commercially, and is the one projected for use by the APL/JHU
OTEC plant discussed in the previous chapter. The main required inputs
are water and electricity. Of course, solar energy indirectly makes the
OTEC electricity possible. Since a multiple-stepped process of energy conversion is prone to energy losses at every step, a more direct approach
would be desirable from both economic and technical points of view.

This is the important aspect of Fujishima and Honda's device. Photo-electrolysis, in which light falling on an electrode composed of a photo-active semiconductor material (such as TiO_2) causes the necessary potential difference and electric current for the electrolytic process to proceed, is an extremely desirable goal. If the photoactivity is sufficient, this kind of cell (Figure 8.3) can produce not only electrolysis, but excess electric power as well. It can be seen, then, that the photoelectrolytic production of fuel and electricity would demand reasonably high efficiencies of operation. The cell operated by Fujishima and Honda achieved a conversion efficiency (light into H_2 and electricity) of about 1%.[102]

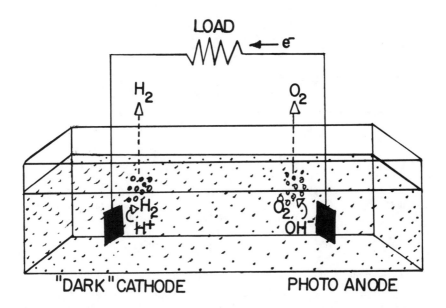

Figure 8.3. The only major difference between this photoelectrolysis cell and that shown in Figure 8.2 is the fact that the process is powered by light input rather than auxiliary electrical input. This is the main charm of photoelectrolysis: that solar energy may be utilized to entirely drive the dissociation process.

The reaction of the photoelectrolytic cell to the incidence of light is roughly analogous to, and indeed is a special case of, the photovoltaic effect associated with solid-state solar cells. The photoelectrolytic cell, then, can be conceptualized as an n-electrolyte-p (n-p photovoltaics were discussed generally in Chapter 5).[103] As is the case with any semiconductor device, the light incident on the electrode must contain sufficient energy to allow the electrons in the semiconductor material to break free of their bonds. This "photoexcitation" creates freed electrons and holes, which subsequently drive the electrochemical redox reactions that occur at the two electrodes.[104] If the incident energy is too weak, nothing happens. The energy level diagram for this kind of device can be seen as Figure 8.4.

P-N JUNCTION ANALOG

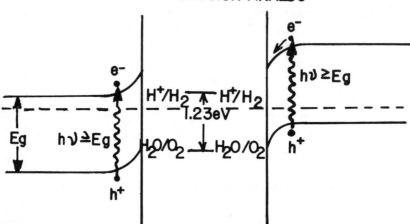

Figure 8.4. The energy level scheme used in conceptualization of the p-n type photoelectrolysis cell. Eg refers to the band gap energy necessary to cause semiconductor electrons to jump from the valence band to the conduction band of the material. "$h\upsilon$" refers to incoming photon energy, and, of course, the $h\upsilon$ energy must be equal to or greater than the necessary Eg for the cell to work.

As is evident, the material used in the electrodes must be matched in the gap energy it needs with the energies available in sunlight. Ideally, the device could make efficient use of most of the solar spectrum that reaches us, dictating the use of a material with a small "band gap" (activation energy requirement). At this point, we come to the prime impediment to the early development of a solar photoelectrolysis cell, which is caused by "a vexing dichotomy in the properties of semiconductor electrode materials."[100] To date, those materials that require the (desirably) lowest band gap energies are also the electrode materials most prone to corrosion or dissolution during photoelectrolysis. Electrode instability is thus the villain that must be exorcised. As stated by researchers at Battelle Columbus, the materials that are "potentially useful semiconductors reported to date which are highly colored, and therefore absorb solar energy effectively, tend to be unstable...under photoelectrolysis conditions, while those materials which have been found to be intrinsically stable also tend to be transparent, or nearly so, to most of the (terrestrial) solar spectrum."[100]

As per A. J. Nozik of Allied Chemical's Materials Research Center, only one type of p-n photoelectrolysis cell has been tested that has both stable electrodes and the capability to support sunlight-induced photoelectrolysis without the additional input bias of an external source. This cell has n-TiO$_2$ and p-GaP electrodes, and has thus far exhibited an efficiency of 0.25% (the Fujishima-Honda cell needed an externally applied voltage bias, and used only high-intensity ultraviolet light for photoexcitation). A partial listing of other experimentally studied combinations is presented in Table 8.1.[105] Obviously, electrodes having good conversion efficiencies are fairly useless if they have a short effective lifetime. On the other side of the coin, stable electrodes having very low efficiencies would also be an exercise in wasteful manufacture.

Putting the situation into perspective, however, it must be remembered that the photoelectrochemical branch of solar research is really only about nine years old, and has suffered from a lack of organization and manpower (as compared with solid-state solar cells, for instance). The field is thus extremely young, with massive room for improvement. In principle, photoelectrochemical cells are capable of demonstrating a number of advantages over more conventional solar cells. The double output of these cells (fuel and electricity) may be perfect for certain applications, and, more importantly, could be varied as necessary from "100 percent electrical (output) to 100 percent storable chemical fuels."[101] In fact, work has been undertaken by Battelle Columbus Laboratories to suppress the photoelectrolytic production of oxygen in operating cells (oxygen production being the leading case of electrode instability). Figure 8.5 shows a schematic of the apparatus they used for the experimentation. Using a light source that closely approximates terrestrial sunlight, a strontium titanate (n-SrTiO$_3$) photoanode, a platinized platinum foil cathode, and an electrolyte solution of 0.5 N sodium acetate in 6 N sodium hydroxide, oxygen production was essentially eliminated with no appreciable change in the cell's current-voltage behavior. Thus, the cell was able to photoelectrolize producing a full complement of hydrogen but no concurrent oxygen, greatly increasing the stability of the SrTiO$_3$ electrode (and thus the life of the cell). Although a large-scale photoelectrolysis plant must, of course, be an open-cycle arrangement to allow for the continuous throughput of water, and, therefore, not privy to an electrolyte composition like the one used in the experiment, the results of the work may point the way toward stable and efficient electrodes for these cells.[100]

Fully developed photoelectrolysis cells will also be able to avoid some of the major drawbacks inherent to solid solar cells (at their present level

Table 8.1. Summary of Experimental Studies on Photoelectrolysis Cells[105]

Cell Configuration	Semiconductor Band Gap (cV)	Electrode form	Electrode stability	Bias for H_2 Evolution (V)
n-Tio$_2$/Pt	3.0	Single crystal, thin film	Stable	0.3 - 0.5
n-SrTiO$_3$/Pt	3.4	Single crystal	Stable	0
n-BaTiO$_3$/Pt	3.3	Single crystal	Stable	0
n-FeTiO$_3$/Pt	2.8	Polycrystalline	Stable	0
n-CaTiO$_3$/Pt	3.4	Single crystal	-	0
n-Fe$_2$O$_3$/Pt	2.2	Single crystal, thin film	Stable with pH > 4.5	0.5
n-WO$_3$/Pt	2.7	Single crystal, polycrystalline	Stable with pH < 7	0.6 - 1.0
n-SnO$_2$/Pt	3.5	Single crystal	Stable	0.5
n-KTaO$_3$/Pt	3.5	Single crystal	Stable	0
n-GaP/Pt	2.25	Single crystal	Unstable	0
n-GaAs/Pt	1.35	Single crystal	Unstable	0
p-GaP/Pt	2.25	Single crystal	Stable	1.0
p-Si/Pt	1.12	Single crystal	-	-
n-TiO$_2$/p-GaP	3.0/2.25	Single crystal	Stable	0
n-GaP/p → GaP	2.25	Single crystal	Unstable/Stable	0

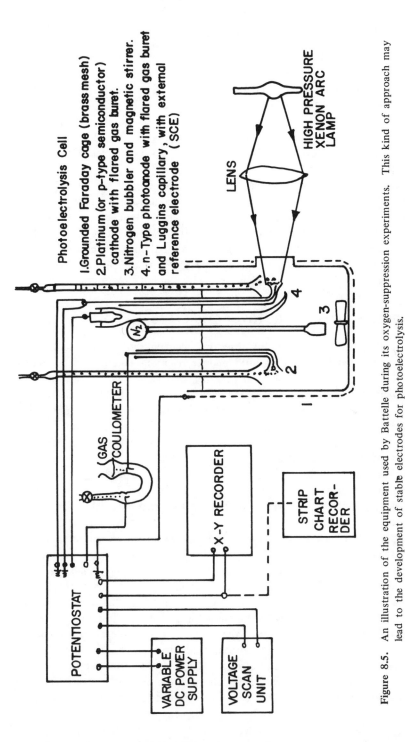

Figure 8.5. An illustration of the equipment used by Battelle during its oxygen-suppression experiments. This kind of approach may lead to the development of stable electrodes for photoelectrolysis.

of development). Even in their present primitive form, photoelectrolysis cells are not as adversely sensitive to polycrystalline semiconductors as solid solar cells, and this one factor should represent a large reduction in cost. There would be no need for antireflection coatings, since the electrolyte would serve that purpose. Cell breakage caused by differences in coefficients of thermal expansion will not exist. All in all, the procedures behind manufacture of future photoelectrolysis cells should prove to be much less elaborate than solid solar cells, once suitable electrode materials are found.

The similarities between the operation of solid solar cells and photoelectrolysis cells should not be allowed to lead to confusion regarding function, however. Photoelectrolysis is primarily intended to produce fuel (hydrogen) as an output, with excess electricity being more a by-product than a purpose (this flexibility may be useful, though, to certain applications). If solid-state photovoltaic cells receive any competition from photoelectrochemical devices, it will be given by "liquid" photovoltaic cells.

ELECTROCHEMICAL PHOTOVOLTAIC CELLS

In these devices, any redox reactions occurring within the electrolyte are neutralized by matching but opposite reactions taking place elsewhere in the liquid. This results in no net chemical change within the cell. The configuration of the devices is analogous to "a Schottky-type cell with one electrode being either an n- or p-type semiconductor, and the counter-electrode being a metal...(the) p-n configuration offers no advantage here because one wishes to maximize the photovoltage, and not the potential between minority carriers as in photoelectrolysis cells."[105] The "minority carriers" mentioned here refer to the fact that, in a heterogeneous p-n cell, the net potential energy available during operation for doing chemical work (production of fuel) is greater than the energy available for electrical work. For this reason, more fuel is produced than electricity. Obviously, a photovoltaic cell should be designed to maximize its capability for doing electrical work.

The input solar energy (light) entering a semiconductor-electrolyte, photovoltaic cell frees some of the electrons within the surface of the semiconductor electrode, creating electron-hole pairs (this behavior is exhibited by any photoactive semiconductor material). After the pair is created, the hole moves toward the electrolyte solution and the electron moves "through the bulk of the semiconductor and then through an external load to the counter-electrode."[106] Of course, this electric current performs electrical work in passing through the external circuit. This work is the effective output of the cell. The electrode reactions that take place in these cells shuttle the electrical charges between the electrodes (rather than reducing and oxidizing the electrolyte), thus completing the circuit.

Various electrode materials have been tested for use in liquid junction solar cells, including both single crystal and polycrystalline semiconductors, and it has been found that the polycrystalline materials (as in the case in photoelectrolysis cells) cause substantially less of an efficiency loss, as compared to the performance of single crystal electrodes, than would have been predicted by work on the more conventional photovoltaic devices.[106,107] Although this is in seeming contradiction to the observation that the "optimum operating characteristics of the electrochemical photovoltaic cell are completely analogous to solid-state photovoltaic devices,"[105] further information demonstrates consistency. In solid-state solar cells, you may remember, the incoming light must pass through the illuminated semiconductor layer to the vicinity of the junction (or barrier), losing energy in the process. In liquid junction solar cells, however, many electrolytes are transparent to significant portions of the solar spectrum, and this mechanism for energy loss is thus eliminated. Morever, the occurrence of imperfect phase matching or dislocations along the junction are all but removed due to a liquid's capability to conform to the electrode's surface.[108,109]

Electrode instability is a problem here as well, and is again due to the tendency for the electrode material to oxidize. The most promising approach to this problem so far has been to engineer the composition of the electrolyte such that harmless (extraneous) reactions are more likely to occur than the electrode corrosion reactions. While encouraging results have been obtained, much more work will certainly be necessary. After extensive development, then, the electrochemical photovoltaic cell could very possibly output electrical power, to be handled in similar fashion to that output by solid solar cells or wind turbines.

As is evident and has been pointed out, the field of photoelectrochemical devices is still very young, and is, in fact, the least developed of the solar alternatives. The field represents a true technological frontier, being less closely related to conventional technologies than the others considered. The promise offered is substantial, but present knowlege is very incomplete. As discussed, the two main types of devices in the field are basically very similar, but are "tuned" to behave differently. An outgrowth of conventional electrolysis for fuel production, this infant branch of research lacks the experience necessary to foresee (even generally) what form the developed devices will take, or for which specific applications they will be best suited. More experience does exist, however, in dealing with the hydrogen end product.

HYDROGEN UTILIZATION

The fuel output from photoelectrolyte cells will be in the form of gaseous hydrogen. The advantages of storing energy as chemical potential have been discussed, and, in hydrogen's case, there are three fundamental ways in which to do it. The densities of hydrogen's energy in these storage systems are listed in Table 8.2.

Table 8.2. Estimated Hydrogen Storage Comparisons[111]

	kWh/lb	kWh/ft^3
Hydrogen Gas (2000 psi)	108.5	10.5
Liquid Hydrogen	11720.0	63.0
Iron Titanium Hydride	240.0	48.8
Magnesium Nickel Hydride	977.0	50.5

H_2, in its gaseous form, can most economically be stored at high pressures (2,000 psi or so) in either aboveground storage facilities or in large underground reservoirs. Aboveground tanks, at these pressures, would be necessarily large and thickwalled massive structures, so that energy compactness will exact a price in facility construction. A considerable amount of electrical energy input will be required to both pump the gas and to compress it, and it should be kept in mind that hydrogen gas carries only about one-third the energy content per cubic foot as natural gas. The underground storage alternative can be carried out (as is natural gas storage at present) in locations such as caverns, aquifers or depleted petroleum reservoirs. Estimates have it that underground storage of H_2 would cost approximately $1-3/million stored Btu. If the cell costs were low enough, it is possible that gaseous H_2 storage of this sort would be practical.[103,110-112]

Another way to compact the fuel would be to cool it to its liquid phase (at temperatures of about -400°F). This approach would put hydrogen into its most energy-dense form (Table 8.2), both on a weight and volumetric basis. Storage in this energy-dense form would also require an extremely energy-intensive methodology, however. It is has been estimated that cryogenic liquification of H_2 would require approximately four times the energy input that the previously described gaseous compression would need.[112] In addition, this process would require highly sophisticated vessels and continuous refrigeration, making liquid H_2 storage the least attractive course of action.

The remaining storage alternative is the residence of H_2 within metallic hydride compounds (iron titanium hydride or magnesium nickel hydride).

For example, FeTi can absorb dehydrated H_2 at room temperature to form the hydride FeTiH. This form would be a convenient, storable form of hydrogen, giving good energy densities as well. To release the hydrogen from the alloy, the hydride would be heated (perhaps with solar energy) to about 200°F. Thus, the "auxiliary" energy input needed for this approach would include transport pumping and the satisfaction of release-heating requirements. Of the three storage methods listed (all admittedly expensive), this latter one seems to give the best energy densities, long-term stability, and economy.[111,112] It must be noted, however, that the inevitable future refinements in storage techniques make predictions of this sort tentative at best.

Once the storage methodology has been determined, the choice of actual usage still remains. We will first omit use of the hydrogen as an industrial chemical feedstock, since the discussion is concerned with hydrogen as an energy carrier. The most straightforward way in which to utilize hydrogen would be achieved by direct combustion of the gas-to-power vehicles or industrial process systems. In fact, research on hydrogen-burning automobiles has already exhibited the excellent mileage characteristics and negligible combustion products associated with the fuel. It has been estimated that hydrogen, as a direct fuel, must attain a cost of about $2-4/ million Btu to be competitive with the likes of No. 2 fuel oil, natural gas, LNG and SNG.[103] This method of usage may also be amenable to the reduction or elimination of the storage component, under certain conditions. Alternatively, H_2 and O_2 could concurrently be fed "into an aphodid burner, and the resulting steam can be used in a steam expander to produce mechanical/electrical energy."[111] This suggests the possible potential of onsite photoelectrolytic processing at large installations.

The probability of an eventual switchover from liquid fossil fuels to synthetic liquid fuels illustrates the values of using H_2 to synthesize alcohols and hydrocarbons. The fuels most widely discussed in this regard include ammonia,[92] methanol, gasoline, ethanol, methane and hydrozone.[111]

The most discussed and, perhaps, the most fitting use of photoelectrolytic output is to feed the hydrogen and oxygen into a fuel cell. A fuel cell similar to the ones used on manned space missions is diagrammed as Figure 8.6. The process taking place inside the cell is the reverse of the photoelectrolytic process. As the hydrogen and oxygen are forced across the electrode's surfaces, they spontaneously undergo charge transfer with the electrodes. The resultant electric current (obtained mainly from electrons given off by the dissociation of hydrogen at "its" electrode) reaches the electrode over which oxygen is bubbling, and the oxygen combines with H^+ ions to produce water. Thus, a general description of the workings of a hybrid photoelectrolysis/fuel cell plant would state that, whereas the former process consumed electricity (from solar energy) to

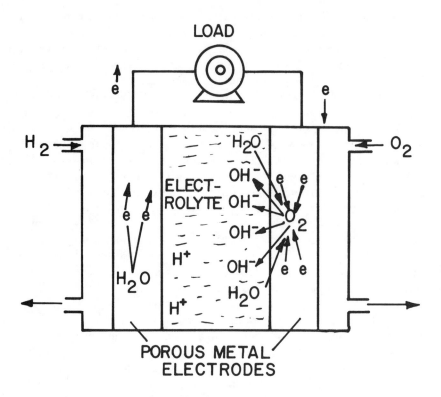

Figure 8.6.   A hydrogen-oxygen fuel cell. This kind of device may prove to be the most beneficial way in which to tap the energy potential of photoelectrolyte output.

split water into H_2 and O_2, the latter process consumed H_2 and O_2 to produce water and electrical energy. The attractiveness of the device, aside from the absence of moving parts, the freedom from pollution and the consumption of capital energy resources, lies in the fact that (unlike thermodynamic cycles) the theoretical efficiency of a fuel cell is 100%. This is due to the absence of any requirement for the "working medium" (moving electrons) to exit from the cell while still retaining most of their energy (as is required in a Carnot heat cycle, for instance). Although this theoretical performance is, of course, never reached, "the real efficiency of the electrochemical energy convertor (a fuel cell) will be more like 60%. . .(compared to internal combustion, 20%)."[113] The technology for fuel cells is reasonably well established, and lends itself, technically, due to a high degree of modularity, to both small- and large-scale power conversion plant scenarios. This gives us, on paper, a choice between centralized and decentralized (even locally owned) utilities.

In this kind of a power system, moreover, storage may be used only as a means to level diurnal or seasonal solar energy availiability.

If the much publicized hydrogen economy does indeed come along, it seems reasonably safe to bet that photoelectrochemical devices will play a large part in its creation and development. Without painting too rosy a picture for the future of this field, suffice it to say that new technologies have been known to blossom within very short periods, and as it stands now, no concepts holding this much promise should be overlooked.

CHAPTER 9

BIOLOGICAL CONVERSION
OF SOLAR ENERGY

INTRODUCTION

The last major remaining approach to tapping the sun's income energy lies within the realm of biology. More specifically, the immensely important biological activity of photosynthetic plants is the central subject. The essential nature of these organisms is largely unrecognized by most people, who tend to think of vegetation only in terms of aesthetics (flowers and landscaping), utility (lumber) and nutrition (fruits and vegetable produce). Other aspects of their significance, in the more fundamental sense, are glossed over. Had photosynthetic plants not evolved, neither would have the earth's oxygen atmosphere and its ozone blanket, since the "primordial atmosphere of the earth contained no free oxygen and was highly transparent to ultraviolet radiation."[114] The continuing impact of this development on our own presence goes without saying.

Even entertaining the fanciful notion that the atmosphere could have come into being through some other mechanism does little to reduce the dependence of all living matter on the photosynthetic process. The process, indeed, constitutes the very foundation of life on this planet, converting sunlight and basic elements into forms of energy more digestible to the "higher" organisms, including man. Thus, although not possessed of a high-efficiency photosynthesis, it is by far the most well established' and spectacularly successful solar energy conversion process in existance. This chapter endeavors to describe this bioconversion process, as well as some of the ways it and related processes may be used to provide even more energy for human purposes.

Whenever photosynthesis takes place, a biological group known as "phototrophs" are using solar energy as the impetus to manufacture the substances they need to maintain and propagate themselves.[115] As a result, their bodies contain chemical energy in the form of carbohydrates and adenosine triphosphate (ATP), a compound basic to all living cells. These substances are, in turn, consumed by "chemotrophs" (including us), who directly benefit from the phototrophic solar-to-chemical conversion. Further information on these "trophic" classifications is presented in Table 9.1.

Table 9.1. The Basic Classification of Life Forms, According to the Ways in Which They Obtain Energy and Carbon[115]

II. Life Forms by the Way They Obtain Energy
    A. Phototrophs (energy from sunlight)
       Algae, higher plants, some bacteria (green, purple)
    B. Chemotrophs (energy from chemicals)
       1. Lithotrophs (inorganic chemicals)
         some bacteria
       2. Organotrophs (organic chemicals)
         Animals, fungi, most bacteria

OR

II. Life Forms by the Way They Obtain Energy and Carbon
    A. Autotrophs (energy from nonliving sources, carbon from CO_2)
       1. Photosynthetic (energy from light)..."A" above
       2. Chemosynthetic (energy from inorganic chemicals)..."B1" above
    B. Heterotrophs (energy from carbon from organic chemicals)..."B2" above

The Process itself is not, as yet, fully understood, due to its complexity. In general terms, chlorophyll molecules within the plant absorb energy from both the visible and ultraviolet portions of the terrestrial sunlight spectrum. The chlorophyll acts as a catalyst in a process by which carbon dioxide and water are split and combined into carbohydrate and oxygen, as in the following formula[116]:

$$CO_2 + H_2O \xrightarrow[\text{chlorophyll}]{\text{solar energy}} CH_2O + O_2$$

While other forms of photosynthetic conversion are carried out by certain algaes and various kinds of bacteria (these reactions require either a gaseous hydrogen source or an organic material), the reaction listed above (representative of green algae and higher plants) is easily the most widely occurring one.

Thus, do green plants generate new and maintain existing tissue (biomass), perpetuating life on the plant. Strictly biomass is a term that has been used to describe both animal- and plant-produced mass, but will be used here to denote only the latter material. Biomass, among other things, is nature's way of storing solar energy, and is produced in more than one form. "Sugar, which is available immediately for energy, is produced in fruits and plants. For short-term storage, starch is produced, which can be converted with little difficulty to fuels. Over a longer period, cellulose is produced, providing long-term storage of energy."[67] All these types of biomass have been and are being used extensively for their chemical (food) and structural (buildings, furniture, paper, etc.) properties, as well as their Btu content (firewood). Further processing, however, can be used to refine this fuel potential, allowing for biomass concerned to be used as a feedstock for the production of upgraded liquid and gaseous fuels. Keep in mind that, as was the case in discussing hydrogen utilization in Chapter 8, only the uses of biomass concerned with energy supply will be described. The applicability of any of the following methodologies to other purposes was not considered directly pertinent to this discussion.

PRODUCTION OF PHOTOSYNTHETIC BIOMASS

Whatever its end use, obviously a key component of any biomass energy system will be production of the biomass itself. Man-induced biomass production at present is logically tailored towards maximum production of end product. In the case of agriculture, crop handling and breeding have traditionally been intended, for instance, to increase the size and number of beans, rather than the size of the whole plant (roots, stalk, leaves, etc). In biomass production, on the other hand, all efforts would be directed toward obtaining the greatest amount possible of biomass per unit input energy per season.

Land-Based Production

The next step is to determine which plant species would be best suited to biomass production optimization. Among the crops thus far deemed suitable for further investigations are sugarcane, sugar beets, sorghums, corn, wheat, forage, grasses, kenaf, eucalyptus, short rotation hardwoods, sunflowers and comfrey.[115,117-119] Many of these crops are currently under cultivation as salable food or feed commodities, and their harvest and/or processing leaves behind the unwanted part of the plant, as biomass residue (currently considered as waste). This residue, as we will see, can be useful. Agronomic practice could be modified to find the optimum farming technique for the coproduction of commodity and residue.

For sugarcane, for instance, current practice dictates that "irrigation water is shut off at appropriate times in the ...fields to ripen the sugarcane. This process stops vegetative growth and maximizes sucrose production."[119] It is felt that a coproduction trade-off of growth technique can be found. For a number of reasons, sugarcane holds the number one position, at present, as a crop for conversion to biomass fuel. Its harvesting season is relatively long. Sugarcane offers the highest yield of any present commercial crop, making it the most efficient photosynthetic converter of solar-to-biomass under consideration. Due to the plant's nature, its biomass residue must already be brought to the sugarcane processing facility, and, thus, no further collection or transport would be necessary for the residue. Moreover, sugarcane is not prone to removing large quantities of nitrogen from its soil, allowing for a lesser use of fertilizer.[119] It has been estimated that an acre of suitable ground devoted to sugarcane production will produce some 20 ton/yr of dry material.

Sugar beets are the next most prolific crop. Although the sugar beet has a shorter harvest season than sugarcane, the harvested product can be stored over longer priods. This plant would be better suited for use as either edible commodity or as biomass (one or the other), since a separate collection system would be necessary to gather its residue for use as the energy source. Further, this residue is nitrogen-rich, making its replacement with fertilizer a necessary adjunct to residue collection, further increasing costs. Moreover, highly cost-effective uses have already been found for leftover sugar beet pulp (as animal feed, primarily), making it more difficult to economically use the material as energy-source biomass.[119] Its fast growth is a big advantage, however.

Grain sorghum is able to thrive on poorer-quality land than can many other crops, a fact that widens the choice of areas in which it can be produced. Its grain is similar to corn as a source of carbohydrate. In addition, the plant is "generally more heat and drought-resistant than corn and requires less water for normal growth."[115] Sweet sorghum is a substantially different plant, being a potentially good source of table sugar, but shares with grain sorghum its hardy nature. Sweet sorghum is also amenable to much more temperate climates than sugarcane, suggesting future market possibilities. Both kinds of sorghum are perennials, and the roots left over from harvested plants are capable of producing shoots for additional biomass production (Figure 9.1).

Corn, as well, could be a profitable candidate for coproduction scenarios. According to E. Lipinsky of Battelle Columbus Laboratories, corn would be a prolific source of both biomass and fermentable sugars. It can be easily stored, a fact attested to by many years of experience. High-quality protein and oils are also to be obtained, and have indeed been exploited on a

Figure 9.1. Scientists evaluating the feasibility of double-dropping sweet sorghum (fuel crop) and winter wheat (food crop) in the midwestern corn belt. (Photo courtesy of Battelle Columbus Laboratories).

large scale. The main modification required for coproduction would be the rebreeding of varities of corn that produce a higher percentage of stalks, leaves and tops, unlike the currently used hybrids (bred to produce maximum food and minimum residual biomass).[119]

The above crops are the frontrunners for coproduction situations, and thus would be the most likely to be able to overcome a major dilemma inherent to organized biomass production. The dilemma is posed by the current worldwide food situation. If high biomass-yielding crops are planted on good agricultural land solely to produce biomass fuel, the food-growing potential of that land is taken away. This kind of agricultural practice would be increasingly hard to reconcile as the world's food requirements grow with the years. Use of coproduction techniques, however, would contribute to both nutritive and energy needs, allowing for synergistic use of farmland (an ever more precious commodity). This approach could also result in an added income to farmers, a significant factor in the concept's economic implementability.

Other crops would seem to be better suited to the production of biomass solely for energy purposes. This approach is the increasingly well-known "energy plantation" concept, in which the Btu content of biomass is all important. The idea is to literally grow fuel for sale as fuel. Various estimates of the cost of this kind of methodology have been generated, one of which came forth with a figure of between $13 and $16 per dry ton of biomass[117] (in 1975 dollars).

A silviculture (forest) energy plantation would most likely utilize fast-growing (short-rotation) hardwoods as the biomass production mechanism. Dense stands of 20,000-25,000 trees/ac could be harvested every four to eight years, depending on which species of tree is used. The most desirable species would be developed from existing varieties already in possession of some of the desired characteristics. These characteristics include rapid growth, resistance to insects and disease (single-crops stands, "monocultures," are notoriously susceptible to diseases and pests),[115] low water requirement, low fertilizer requirement, high energy content and amenability to coppicing (the regrowth of a planting after harvest by sprouts or root suckers rather than seed). It has been stated that present expertise should allow for the production of some 7 dry ton/ac/yr.[118] This, of course, would increase considerably as knowledge advances.

A cord of hardwood (equivalent to a stack 4 ft high, 4 ft wide, and 8 ft long) can possibly contain some 25 million Btu of energy potential. Of course, the hardwoods contain much more energy than the softwoods (twice as much), and less than "good" coal (about half as much).[67] The most likely varieties of trees at this early stage of research include sycamore, poplar, ash, sweetgum, alder and eucalyptus.

Such are the main ways in which biomass can be obtained from the land. Alternate methods seem feasible, however, that would use water as the medium of growth. In fact, the sheer size of our world's water-covered surface indicates that the possible yield may exceed the land's biomass potential.

Water-Based Production

One recent development promises to be useful in two important spheres of human concern: energy and water purification. Intermittent research over the past 30 years has shown the prodigious capability of the water hyacinth to assimilate large amounts of pollutants from contaminated water. In keeping with demonstrated high removals of nitrogen and phosphorus (the substances leading to the eutrophication of lakes and waterways), water hyacinths have been shown to be a feasible method of "polishing" the effluent from wastewater treatment plants. As the hyacinths grow, they tend to fully cover the available water surface in dense stands, yielding reported densities of 125-184 wet tons of biomass per acre.[120] Further, more research has shown that this prolific plant is quite amenable to anaerobic digestion after harvest, a biomass conversion technique that will be discussed.

Another water-bred phototroph that offers double-edged potential is microalgae; simple microscopic plants with a short lifespan and a considerably high rate of reproduction. In 1955, A.E. Fisher of Arthur D. Little, Inc. reported that algae production could be carried out quite feasibly at rates of some 35-50 tons dry weight/ac/yr.[121] The algae were to be grown in shallow, waterfilled tanks or ponds exposed to the sun. Since that time, laboratory experiments have achieved yields of 75 tons dry weight/ac/yr.[3]

Algae are simple creatures, requiring only carbon dioxide, inorganic nutrients and sunlight to thrive. A green algae by the name of Chlorella pyredenoisa has an energy content of approximately 13,000 Btu/lb, rivaling that of bituminous coal.[67] Their protein content (50-62%) is quite high also, again indicating a crop with food/fuel potential. The solar-to-biomass conversion efficiency of Chlorella may eventually be coaxed to reach as high as 20%, although accomplished cultivation thus far has achieved more modest efficiencies of some 10% or so. This, however, is still a good deal higher than most land-based crops. Chlorella, in conjunction with water hyacinths, pretty much exhausts the alternatives available for biomass production on landlocked freshwater surfaces. But then the oceans are still there.

The oceanic version of energy plantations involves the cultivation of giant kelp, a species of brown seaweed. H. Wilcox of the Naval Undersea Center has proposed the large-scale growth of this plant as a source of biomass for chemicals and animal feed, as well as fuel.[122] A 1,000-acre marine farm research project has been proposed, at an estimated cost of $48 million. A full-scale farm would be sized at about 100,000 acres and would cost on the order of $2 billion. United Aircraft Research Laboratories is also involved in investigating large-scale kelp cultivation. Their concept entails seeding 80-mile strips of ocean with kelp, and estimates predict a dry weight yield of about 1 ton/ac/yr. While this cannot be considered a high unit yield, the enormous size of a working kelp farm would tend to overshadow that fact with a huge total production. This situation is analogous to the OTEC conversion system discussed in Chapter 7. Although the energy conversion efficiency was fairly low, the huge quantities of heat from "free" seawater made the process not only feasible, but competitive.

SOLID AND ANIMAL WASTES AS BIOMASS

Primary (first generation) production of biomass is not the only source that can be exploited. Convertible energy is also contained within municipal and industrial solid waste (garbage, refuse, trash, packaging, etc.) and animal waste (manure). The quantities of these "wastes" generated at present are quite large, and the energy that may be gleaned from them is significant. As technology advances (as well as the need for energy), these materials will increasingly be thought of as resources rather than wastes. Thoughts and actions will turn to recovery and utilization instead of simple disposal. As with most of the other resources discussed, these materials hold potential as both recoverable materials and recoverable energy. Further experience will show the relative value of each, but the energy potential is what concerns us here.

Systems already exist nationwide for the collection and transport of solid waste (not to be confused with human excreta) to central points, at which the vast majority is buried in a sanitary landfill or combusted in an incinerator (almost none of these incinerators have the capability to use the energy released by the combustion—it simply goes up the stack). These forms of solid waste management are indicative of the traditional concept of solid waste as useless detritus. By and large, none of this material is recycled for subsequent reuse. As a matter of fact, "average" municipal solid waste contains some 10 million Btu/ton of releasable energy. The methods being developed to tap this resource will be discussed later in this book.

Manure from animal feedlots also represents a considerable amount of releasable energy. From the extremely large resident population of domesticated cows, horses, pigs, chickens and other animals in this country, it has been estimated that more than 2 billion tons of manure are produced each year. Through bacterial degradation, the energy equivalent of about 1 gallon of gasoline can be produced from each 225 lb of manure.[67] In these terms, the energy resource contained within this "waste" makes itself evident.

This chapter thus far has touched on the various sources of biomass that are or can be made available for energy conversion. Many of the methodologies that will be described have been used for years for similar or related applications. Some have been adapted recently in the laboratory for applicability to biomass. The rest have not as yet been operated onto an appreciable scale. All hold some promise, however, and are pertinent to this subject.

CONVERSION OF BIOMASS TO ENERGY

While the different forms of biomass all fall within the category of fixed-carbon-bearing materials, various specific materials respond best to conversion by certain types of processing. The optimum process to be used depends on the material involved, its designated economic value, the region in which it was obtained, and the desired end product.

Direct Combustion

As always, the simple burning of a material is the most straightforward way in which to release its energy. Biomass is of course no exception. In fact, this is the method that has supplied the vast majority of man's needs for heat throughout history (until recent times).[116] Perhaps surprisingly, direct combustion of cellulosic biomass at present constitutes the single most significant use of (indirect) solar energy in this country. Aside from skyrocketing sales of woodburning stoves and the emergence of new interest in fireplaces, industry contributes heavily to the production of biomass energy. "The wood products industry now derives...40 percent of its total energy needs, or about 1 quad (one billion billion Btus) from burning bark and mill wastes."[122] In addition to using the Btus obtained from the biomass as sources of space heat and industrial process steam, it follows that the generation of electrical power would be an application worthy of detailed investigation (this work remains to be done). Table 9.2 serves as an indication of the efficiency of this approach, as presently envisioned.

Essentially, all biomass types mentioned earlier could be combusted directly. Of course, those containing the least amounts of water and the densest dry mass would be most suitable, such as wood and certain of the agricultural residues. Commercial solid waste management systems are becoming available which, as the main end product, process the cellulosic and some of the putresible fraction of solid waste (paper, corrugated board, discarded food) into a "refuse-derived fuel" (RDF), which can be marketed for combustion in fossil-fueled power plant boilers. The extremely low sulfur content of biomass is especially advantageous in this regard, and

Table 9.2. Percentage Efficiency of Electrical Power
Production from Biomass and Coal[117]

	Deep-Mined Coal[123]	Surface-Mined Coal[123]	Biomass
Extraction/Production	57.0%	80.0%	100.0%[a]
Resource recovery	0.8	0.8	5.6
Energy input	——	——	——
Net available	56.2%	79.2%	94.4%
Processing and Transport	b	b	c
Conversion	38.0	38.0	34.1
Transmission	91.2	91.2	91.2
Net delivered	19.5	27.5	29.4

[a]Excludes photosynthetic efficiency.
[b]Mine-mouth plant.
[c]Included in energy input: extraction and production.

simplifies the purification of plant exhaust gases before atmospheric
discharge.

A main drawback of the direct combustion method is due to the relative inconvenience of the use of solid fuels, as opposed to more conventional and consistent liquid or gaseous fuel forms. Thus, conversion of
biomass to these forms must be considered to attain consistency of
product, increased energy densities, and compatibility with existing fueled
process machinery. A number of means to this end are being investigated.

Anaerobic Digestion

The word "decay" refers to a natural process during which bacteria
slowly consume and digest organic material, forming gas and residual
solids as by-products of their metabolism. If the organic material undergoes decay with access to the atmosphere, aerobic digestion (decay in the
presence of air) takes place. The by-products of this process include
ammonia, carbon dioxide and a solid sludge known as humus. More energy-
valuable by-products are produced, however, if air is not available to the
bacteria, in which case anaerobic bacteria (those that thrive in the absence
of air) take over. According to Schellenbach et al.,[124] more than one
group of bacteria are involved in the decomposition process. "The first
group of bacteria breakdown the organic material into simple compounds,
mainly acetic acid (the chief acid in vinegar). The second group convert
the acids produced by the first group into methane gas," carbon dioxide

and traces of hydrogen sulfide. The methane content of the gas varies between 50 and 80% depending on the nature of the organic material.[125]

A considerable amount of experience has been gathered from the use of this process as a means of reducing the sludge output from many sewage treatment plants, in which human feces is removed from wastewater and broken down for purposes of disposal. The main difference between this engineered anaerobic digestion and that occurring in nature is that the process has been greatly accelerated by man, through the addition of heat (up to 95°F, or so) and airtight digestion tanks. The process has also been applied extensively in India to cow manure, to produce methane for various uses.[67]

It has been estimated that as much as 12-33% of an animals's energy intake remains in the manure, "the remainder being apportioned to activity, body maintenance, heat loss and meat, milk or egg production."[125] A company by the name of Bio-Gas of Colorado, Inc. has been working since early 1974 toward utilizing this energy resource. It has developed a system within which the biogas is separated into its methane (CH_4) and carbon dioxide (CO_2) components. The CH_4 may then be sold to local gas companies or used onsite for a number of applications (internal combustion engines, conversion to electricity, grain drying, cooking, etc.)[125,126] The CO_2 is used in this system to support the growth of algae in covered ponds, which "purifies the water (used to form a manure feed slurry) so it can be recycled in the process. Then the excess algae, which is 50-60% protein, is added to the (solid) residue from the digester to form...cake which is refed to the cattle."[126] Figure 9.2 is a simple diagram of the system.

In addition to manure, anaerobic digestion could also be applied to algae, water hyacinths and other suitable biomass types such as kelp, fortunate because of the high water content of these materials (making them ill-suited to direct combustion). This biogasification of organic materials is also being investigated by the University of Illinois, Cornell University, The Institute of Gas Technology and others.[127] It has been estimated that the costs for methane generation through bioconversion would range between \$2.40 and \$2.86/1,000 ft^3, in 1976 dollars.[128] While this price is not, at present, economically competitive with natural gas as a source of fuel, the digestion of fresh manure at a total confinement livestock feedlot would have the additional benefit of reducing pollution of the surrounding area from storm runoff. This added function may possibly make the process economically attractive, but more study is necessary to determine this.[129]

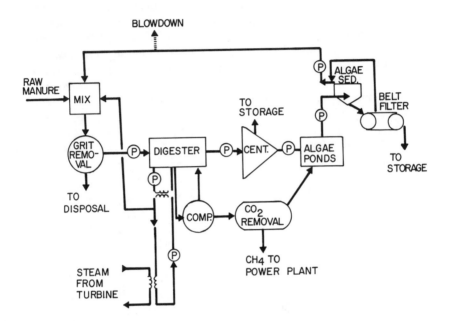

Figure 9.2. A flowchart of the process designed by Bio-Gas of Colorado, Inc. for the digestion of manure from 50,000 feedlot cows. The output gaseous fuel would be combusted at a nearby electric utility.[127]

Fermentation

This technique is also an outgrowth of a natural process, in which the action of enzymes (organic compounds produced by plant or animal cells that acts as a catalyst to reaction) causes a chemical change in a material. The most commercially significant and well-developed fermentation techniques are those used in the production of potable spirits (whiskey, rum, etc.) through the fermentation of field-grown carbohydrates, followed by a distillation process. The method can be applied to convert biomass into ethanol (ethyl alcohol). The biomass types that would lend themselves to this sort of conversion are: sugarcane juice, sugar beet juice or glucose from starchy grains (corn, sorghum). A concept developed by Battelle Columbus entails the use of "fibrous by-products (of sugarcane) to manufacture (through combustion) the steam for processing the agricultural inputs, with enough steam left over to conduct the ethanol distillation."[119]

Preliminary estimates by Battelle place the cost of such ethanol production at about \$1.10/gal in 1976 dollars. For this reason, it appears that the ethanol would not be competitive with gasoline as a fuel, but may be competitive with industrial denatured ethanol, when petroleum costs reach about $20/bbl.[119] The main problem involved with this would not be technical, however. Rather, it is related to the high level of regulatory constraint placed on any kind of alcohol production, especially on a large scale. This condition is so complex that it may well make biomass-to-ethanol conversion effectively impossible to implement.

Pyrolysis

In general terms, pyrolysis is the destructive distillation (at high temperatures) of a material in the absence of air, causing that material to break down into its fractional parts. If the process is applied to wood, the outcoming products are acetic acid, methanol (methyl alcohol also called "wood" alcohol), hydrocarbons and residual charcoal. The methanol, of course, would be collected separately from the other fractions. As a fuel, methanol was half the energy content of gasoline, but burns much more cleanly. Sweden used large-scale production of methanol for fuel during wartime, and it is used at present in clean-air auto races.[67] It is also used in industry "as a solvent and as starting materials for an entire family of industrial chemicals."[119] Of course, this biomass-to-methanol process would be heir to the same alcohol regulations as would ethanol manufacture.

Pyrolysis is also being developed as a means of converting municipal solid waste into useful liquid (120,000 Btu/gal of oil), gaseous (500 Btu/ft^3) and solid (700-1200 Btu/lb char) fuels suitable for subsequent combustion.[115] Process development is not at a competitive stage as yet, due to problems with variations in solid waste composition and the necessity for very large-scale facilities to bring down unit costs. Much more work is necessary before this methodology becomes widespread. The aforementioned RDF processes are, at present, the most successful methods with which to recover solid waste's energy content.[130]

Other Processes

Various other, more exotic, processes are being researched, such as hydrogasification, in which organic wastes and hydrogen are combined at high temperatures and pressures to produce ethane and methane as substitutes for natural gas. Another process, hydrogenation, combines organic wastes with high-temperature steam at high pressure and carbon monoxide to form low-sulfur oil (16,000 Btu/lb).[115] The commercial use of such processes as these in the near future is extremely doubtful, but they do, at least, serve to emphasize the many ways in which the chemical energy locked within biomass can be utilized as fuels to supplement (and perhaps eventually replace) the use of nonrenewable fossil fuels with renewable biomass energy.

As with all new technologies, many technical, economic and political hurdles lie in the path of the widespread adoption of biomass as a major source of energy. It is fairly certain that some of the aforementioned processes will see some site-specific fruition at least, such as combustion and anaerobic digestion. The remaining techniques, however, have a future that is highly speculative.

CHAPTER 10

A LOOK AT THE FUTURE

Just as the human body survives by metabolizing food, so too does human society have an undeniable energy need for its continuance. The realization that capital energy resources are inherently finite and, in fact, largely depletable within generations, has begun to be assimilated, thus spotlighting the need for development of methods with which to harness income energy (inherently undepletable) resources. Throughout history, the release of renewable impounded income energy sustained mankind's need for energy. The ushering in of the petroenergy age changed this situation, however.

Admittedly, the widespread use of petrochemical energy (age-old photo-synthetically impounded solar energy) made possible this century's astounding scientific and social progress. Indeed, it allowed mankind's development to shift into a higher gear, resulting in quantum leaps of interdependent intellectual advancement. This phenomenon exacted its price, however, in both environmental (widespread pollution and scarring of the planet) and philosophical (a materialistic society based on the flagrant use of cheap energy) terms. Another unfortunate outcome of the availability of cheap energy was in economic terms: it made of low economic priority the active research of alternative energies. This is evidenced by the ways in which solar energy has alternately been viewed as both "interesting but of little use" and "fascinating but as yet impracticable," as reflected in countless reports on the subject in past and even very recent years.

These characterizations were nowhere near to giving a comprehensive picture of the actual situation, however. Among other things, there is a self-fulfilling prophesy involved here. Although government is pumping many more dollars into solar energy than at any other time in history, the fact remains that the funding is not even roughly the same as (as a prime example) nuclear energy development. Thinking of this situation in very general terms may help to see the effect of such an arrangement.

Nuclear energy technology, especially state-of-the-art technology (even excluding fusion power), is concerned with extremely complex principles and systems. Immense amounts of cash have been injected into this branch of research since its very inception, almost two generations ago. As a scientific endeavor, the entire space program would be its closest rival (in sheer magnitude). In spite of this enormous economic and manpower input, however, nuclear energy remains today a minor source of energy on a nationwide scale. Heavy controversy rages to this day (some of it sensational, some valid) over very basic problems inherent to the operation of the techniques, including such items as process safety, waste disposal, international security and (ironically) eventual fuel depletion. Whether nuclear energy installations become a major component of the national energy supply, the end product will certainly not be the pristine energy panacea predicted in years past.

The self-fulfilling prophesy aspect now comes to the fore. No one ever (rationally) suggested that atomic energy could become feasible without a substantial, expensive, concerted effort by both the public and private sectors. This same attitude must be taken during the early development stages of all new technologies. If the parties involved directly or indirectly with that development treat it as if it were many years away, it will indeed take many years before it arrives. To a substantial extent, this has been and is still going on with the solar technologies, for a multitude of reasons.

As a group, the solar energy technologies are by no means the greatest scientific challenge ever faced, involving principles and material behaviors intrinsically more manageable than those associated with large-scale sustained nuclear reactions. Most of the principles are essentially extensions of those previously developed for other reasons, and this may be one of the problems. No group or handful of groups have any real prospect of controlling the harvest and distribution of solar energy (as opposed to the nuclear and fossil fuel energy hierarchies). Many of the solar conversion methodologies described herein are best suited to decentralized operation. Even the collection and conversion equipment itself will be manufactured from fairly well-dispersed materials, by any number of manufacturers (large and small). Herein lies solar energy's main strength and most potent weakness.

Among solar energy's most attractive charms are its conceptual simplicity, benign nature, diversity and the intuitively captivating promise of a lessened dependence on the power utilities. In conjunction with the inevitable rise in conventional energy prices, these are the attributes that will swing public sentiment toward active acceptance of solar energy, once cost-effectiveness is established.

And here is the main rub, in which solar energy's diversity will work against its own development. To overcome the enormous momentum that has resulted from many years of cheap energy and its power brokers, the proponents of solar must present a strong and organized front. This unity must be applied to both scientific research and to political and economic implementation. This loosely translates into a need for increased communication between researchers (to avoid wasteful and time-consuming duplication of effort), for better organization in the delivery of information to the public, and for a less fragmented approach to changes in legislation. Unfortunately, this organization will probably come into being only after government funding has served to demonstrate a technology that is (or very close to becoming) economically sound.

Another paradox: cost-effectiveness equipment (especially for individual ownership) will become available only if manufacturers see the presence of a market sufficient to justify their capital investment in mass production systems, and the consumers at large (the potential market) will furnish this demand only after reasonably cheap equipment is on the shelves. This is why the government funding and project grant subsidies are so all-important. They must be carried out to supply the artificial market stimulation necessary to start up the process. Appropriate changes in building codes and tax structures, as well as new legislation regarding the right to direct sunlight must also be brought about to support this development.

This decade is witnessing the beginning of the end of the cheap fossil energy era, symbolically initiated by the handwritten "no gas" signs that sprouted some five years ago. Moreover, this situation must not be thought of as simply using new energy sources to satisfy the present level of energy consumption. The increasingly large chasm that exists between present energy production and virtually all projections of future energy requirements emphasizes the need for a change in attitude. This awakening will be rude and even painful for most people, but the sooner the better. Changes will be felt in all spheres of activity, including architecture, industry, mass and personal transportation, and in individual awareness. Solar energy (and other alternatives) will be able to shoulder a large share of future energy needs much sooner than the year 2000, but only if President Carter's "moral equivalent of war" (not necessarily his proposed energy policy) is indeed realized very soon.

From a long-term perspective, we, as a species, are facing the start of a new stage of development. The slow, hesitant development characteristic of man's history ended with the emergence of cheap fossil energy. That energy brought with it the means for humankind to being a rapid advancement of its technological and intellectual skills, a development that becomes synergistically more rapid each year. The fuel for this process, however is beginning to get scarce, threatening the continuance of progress. We must immediately start to use our collective intellect and technological wherewithal as the tools with which to graduate to a new stage of human development—a stage in which renewable (therefore limitless) resources are used to support further elevations in the worldwide standard of living and other realizations of our potential. If we are wise enough to take this essential step, solar energy will definitely play a major role.

APPENDIX A

Tables A.1 through A.14 are *Insolation on Inclined Surfaces* data and are reproduced with permission from the *1972 ASHRAE Handbook of Fundamentals.*

Included in this appendix section are twelve *Mean Daily Insolation Maps* reprinted from the *HUD Intermediate Minimum Property Standards Supplement,* 1977 edition.

Table A.1. Btu/ft² Insolation on Inclined Surfaces at 24° North Latitude[a]

BTU/FT² Hourly Total Insolation, $I_{D\theta} + I_d$, On Surfaces Tilted With Respect To The Horizontal

Column orientation within each tilt group (AM / PM): E/W, SE/SW, S/S, SW/SE, W/E

Date	AM	PM	0°	14° E	14° SE	14° S	14° SW	14° W	24° E	24° SE	24° S	24° SW	24° W	34° E	34° SE	34° S	34° SW	34° W	44° E	44° SE	44° S	44° SW	44° W	90° E	90° SE	90° S	90° SW	90° W
Jan. 21	7	5	10	25	26	17	4	4	35	36	21	4	4	44	45	25	4	4	52	53	28	4	4	66	68	31	2	2
	8	4	83	128	135	110	68	34	156	167	126	55	13	179	195	137	41	13	197	216	145	25	14	201	230	127	7	7
	9	3	151	193	209	188	143	101	217	242	207	132	61	234	269	221	118	20	244	288	228	100	49	200	262	176	47	8
	10	2	204	233	257	246	208	163	246	286	268	203	128	252	307	282	192	90	250	318	287	178	120	154	252	207	108	9
	11	1	237	249	281	283	255	212	249	302	306	258	186	242	315	319	254	155	227	318	324	247	181	86	217	226	167	9
	12	12	249	242	280	296	280	242	228	292	319	292	228	208	296	332	296	208	182	291	336	291	182	9	167	232	167	9
Feb. 21	7	5	35	71	67	44	16	9	94	88	49	9	9	114	106	53	9	9	131	121	56	8	8	156	141	46	5	5
	8	4	116	167	167	135	90	58	198	198	145	69	15	223	223	150	46	14	241	242	151	22	14	233	234	102	8	8
	9	3	187	212	240	213	168	131	256	269	225	149	87	273	291	230	126	40	281	304	228	100	15	217	249	141	9	9
	10	2	241	272	288	273	235	197	286	312	286	220	158	288	326	291	203	115	284	330	287	178	69	164	231	168	61	10
	11	1	276	288	312	310	284	248	286	317	324	280	220	276	328	328	267	180	258	328	323	247	144	92	192	185	120	10
	12	12	288	279	310	323	310	279	264	316	337	316	264	240	312	341	312	240	210	299	335	299	210	10	138	191	138	10
Mar. 21	7	5	60	104	94	63	30	14	132	115	64	13	13	156	133	62	13	13	175	147	59	12	12	195	154	27	7	7
	8	4	141	193	186	150	107	81	224	212	152	79	36	248	232	149	49	17	265	244	142	33	16	241	212	64	12	12
	9	3	212	256	236	226	185	155	279	279	229	159	109	293	293	225	128	94	302	300	214	94	18	219	218	95	18	16
	10	2	266	296	304	285	251	221	307	320	288	231	181	309	327	283	204	163	302	324	270	172	87	165	197	120	94	18
	11	1	300	312	327	322	299	272	309	334	326	288	242	297	332	320	268	205	276	320	305	240	163	93	156	135	136	18
	12	12	312	303	325	334	325	303	286	323	339	323	286	261	312	333	312	261	228	291	317	291	228	11	102	140	102	11
Apr. 21	6	6	7	16	12	5	4	4	23	16	4	4	4	28	19	4	4	4	33	21	3	3	3	41	24	2	3	3
	7	5	83	127	111	77	45	35	155	127	70	19	19	178	140	62	18	18	196	149	51	17	17	202	134	10	17	17
	8	4	160	209	194	157	120	103	236	210	149	86	59	257	223	137	51	61	270	228	122	85	21	230	169	16	21	21
	9	3	227	259	259	227	192	174	287	272	220	161	130	299	278	206	135	130	302	277	186	157	105	207	148	61	85	23
	10	2	278	287	294	282	254	221	314	310	275	228	197	302	313	259	195	197	303	294	237	220	175	157	110	74	148	105
	11	1	310	307	303	316	299	272	316	313	309	280	254	300	313	293	254	258	280	294	269	267	235	90	60	79	60	175
	12	12	321	312	323	328	323	312	295	313	321	313	295	270	294	305	294	270	235	267	280	267	235	15	15	79	15	15
May 21	6	6	22	41	31	15	10	10	53	36	10	10	10	64	41	9	9	9	73	44	9	9	9	86	43	5	5	5
	7	5	98	140	120	85	57	51	166	131	73	25	24	187	139	59	23	23	202	143	44	21	21	197	112	12	12	12
	8	4	171	215	196	159	127	117	239	207	145	91	75	256	212	127	59	31	267	211	106	26	48	217	135	15	16	15
	9	3	233	270	256	224	194	184	287	263	210	160	142	297	262	190	121	96	299	255	165	79	116	195	114	22	26	17
	10	2	281	305	297	275	252	242	312	298	261	222	205	310	290	239	186	162	299	274	211	145	181	149	78	34	79	48
	11	1	311	319	317	307	294	286	314	301	293	271	258	295	295	270	240	222	277	274	240	203	236	87	32	37	114	116
	12	12	322	313	316	317	316	313	295	304	304	301	295	270	278	281	278	270	236	250	250	250	236	17	17	37	32	17
June 21	6	6	29	50	37	20	13	13	63	42	12	12	12	75	46	12	12	12	84	49	11	11	11	96	45	7	7	7
	7	5	103	143	122	87	61	57	168	131	73	29	26	187	137	58	25	25	201	139	49	23	23	191	101	13	13	13
	8	4	173	215	194	158	128	122	238	203	142	92	81	254	206	122	59	39	263	203	95	44	55	209	120	16	16	16
	9	3	234	268	252	221	186	186	284	258	204	158	146	292	254	182	119	101	291	244	155	76	119	188	99	18	18	18
	10	2	280	302	294	269	242	233	308	290	253	218	206	305	280	229	181	165	294	263	199	139	181	144	65	19	55	19
	11	1	309	316	311	300	284	284	311	303	290	270	257	296	285	259	233	222	273	263	227	194	235	85	21	22	101	21
	12	12	319	310	310	310	310	310	293	294	294	294	293	267	269	269	269	267	235	236	236	236	235	19	21	21	21	19

BTU/FT² Hourly Total Insolation, $I_{D\theta} + I_d$, On Surfaces Tilted With Respect To The Horizontal

Date	Solar Time AM	Solar Time PM	0°	14° E/W	14° SE/SW	14° S	14° SW/SE	14° W/E	24° E/W	24° SE/SW	24° S	24° SW/SE	24° W/E	34° E/W	34° SE/SW	34° S	34° SW/SE	34° W/E	44° E/W	44° SE/SW	44° S	44° SW/SE	44° W/E	90° E/W	90° SE/SW	90° S	90° SW/SE	90° W/E
July 21	6	6	23	40	31	16	11	11	52	35	11	11	11	62	39	10	10	10	70	42	9	9	9	81	41	6	6	6
	7	5	98	138	118	85	57	52	162	128	73	57	25	182	136	59	24	24	196	139	44	23	23	190	107	13	13	13
	8	4	169	212	192	157	126	118	235	203	143	91	77	252	207	125	54	35	261	206	104	28	28	210	131	18	16	16
	9	3	231	266	251	221	192	182	282	258	207	159	142	291	257	187	120	98	291	249	161	79	51	190	130	21	18	18
	10	2	278	301	292	270	248	239	307	292	256	219	203	304	285	235	184	162	293	269	206	143	116	146	110	32	17	17
	11	1	307	315	312	302	289	282	310	305	287	266	255	295	289	265	236	220	273	266	233	199	179	86	76	36	17	17
	12	12	317	308	311	312	311	308	291	296	298	296	291	266	273	275	273	266	233	242	245	242	233	19	31	58	31	19
Aug. 21	6	6	7	15	12	7	4	4	20	15	4	4	4	25	17	4	4	4	29	21	4	4	4	36	21	2	2	2
	7	5	82	122	107	76	47	37	147	122	69	22	22	168	133	60	21	27	184	140	53	25	25	187	124	11	11	11
	8	4	158	203	188	154	119	104	228	204	146	87	130	290	269	200	53	85	292	218	124	86	138	218	159	16	15	15
	9	3	223	261	252	222	189	172	279	264	214	159	195	290	299	252	124	153	295	266	192	155	204	199	162	39	16	16
	10	2	273	298	295	275	249	232	306	302	268	224	250	295	305	285	198	215	274	288	261	215	220	153	142	58	17	17
	11	1	304	313	317	309	292	278	309	315	301	274	289	264	287	296	248	264	231	285	261	260	266	89	106	71	17	17
	12	12	315	306	316	320	316	306	289	303	313	306	256	287	287	296	287	266	287	260	272	272	233	17	58	75	58	17
Sept. 21	7	5	57	96	87	60	30	16	120	106	60	15	15	142	122	59	15	15	159	134	56	14	14	175	139	26	8	8
	8	4	136	184	178	144	104	80	212	201	146	78	38	235	219	143	50	21	250	231	136	20	20	226	199	62	11	11
	9	3	205	247	246	218	179	152	268	268	221	155	108	282	281	217	126	62	287	287	206	94	22	209	208	93	13	13
	10	2	258	286	293	275	243	215	296	308	278	224	177	298	314	273	199	134	291	311	261	168	160	159	189	116	13	13
	11	1	291	301	316	309	290	264	298	313	315	279	236	287	321	309	259	200	267	309	295	233	221	91	151	131	42	13
	12	12	302	294	315	323	315	294	277	313	327	313	277	253	301	321	301	253	221	281	306	281	204	14	100	136	100	14
Oct. 21	7	5	32	6	60	40	15	10	83	78	45	10	10	100	94	48	9	9	115	107	50	9	9	136	121	42	5	5
	8	4	111	158	159	129	87	51	187	188	139	67	17	211	212	144	46	16	228	230	145	23	13	219	221	99	23	9
	9	3	180	223	231	206	163	127	246	259	217	145	85	262	280	223	124	41	272	292	279	98	50	208	240	138	98	10
	10	2	234	263	278	265	228	191	275	303	277	216	154	279	315	282	198	113	274	320	314	174	120	159	225	165	174	11
	11	1	268	279	303	301	278	241	278	317	315	272	214	268	323	319	260	180	250	318	320	241	181	90	188	185	76	76
	12	12	279	271	301	314	301	271	256	307	328	307	256	233	303	332	303	233	204	291	327	291	204	11	136	188	136	164
Nov. 21	7	5	10	24	25	16	4	4	34	34	20	4	4	42	43	24	4	4	50	51	27	4	4	63	64	29	4	1
	8	4	82	126	133	108	68	34	153	164	123	55	14	175	190	135	41	13	193	212	142	26	13	197	223	124	7	7
	9	3	150	191	206	186	142	101	214	239	205	131	62	231	265	217	117	21	241	283	224	99	50	196	257	172	9	9
	10	2	203	231	255	244	206	163	244	283	265	201	128	249	303	273	190	90	247	314	283	174	120	152	247	204	46	10
	11	1	236	247	278	280	252	211	247	299	302	255	185	240	311	316	251	155	225	314	320	239	181	86	213	222	106	10
	12	12	247	240	277	293	277	240	227	315	315	289	227	207	293	328	293	207	181	288	332	288	181	10	164	228	164	10
Dec. 21	7	5	3	10	10	7	2	2	14	15	9	2	2	18	19	11	2	2	21	22	12	1	1	27	29	14	1	1
	8	4	71	112	121	99	59	26	138	152	116	49	12	160	179	129	38	12	177	201	139	26	11	185	220	130	6	6
	9	3	137	177	195	176	132	89	200	229	198	124	57	217	257	214	113	41	228	277	223	98	41	190	262	184	8	8
	10	2	189	217	243	234	195	150	230	273	258	193	116	236	296	275	185	110	235	309	283	172	110	148	247	217	58	9
	11	1	221	233	267	270	241	197	233	290	295	247	173	227	304	312	245	170	213	310	320	237	170	84	222	236	117	9
	12	12	232	226	266	282	266	226	213	280	308	280	213	194	287	325	287	194	170	284	332	284	170	9	174	243	174	9

aBased on data in Table 1, p. 287, 1972 *ASHRAE Handbook of Fundamentals*; 0% ground reflectance, 1.0 clearness factor.

Table A.2. Btu/ft² Insolation on Inclined Surfaces at 32° North Latitude [a]

BTU/ft² Hourly Total Insolation, $I_{D\theta} + I_d$, On Surfaces Tilted With Respect To The Horizontal

Date	Solar Time (AM/PM)	0°	22° E/W	22° SE/SW	22° S	22° SW/SE	22° W/E	32° E/W	32° SE/SW	32° S	32° SW/SE	32° W/E	42° E/W	42° SE/SW	42° S	42° SW/SE	42° W/E	52° E/W	52° SE/SW	52° S	52° SW/SE	52° W/E	90° E/W	90° SE/SW	90° S	90° SW/SE	90° W/E
Jan. 21	7/5	0	1	1	0	0	0	1	1	0	0	0	1	1	0	0	0	1	1	1	0	0	1	1	1	0	0
	8/4	56	114	125	93	37	56	136	151	106	37	56	153	175	116	37	56	167	190	123	10	10	171	200	115	6	6
	9/3	118	177	203	175	104	104	196	233	193	99	99	210	256	206	99	87	217	272	274	73	14	187	256	181	8	8
	10/2	167	208	249	235	175	156	217	275	256	171	162	219	293	269	175	148	216	302	295	148	67	147	257	221	61	9
	11/1	198	212	267	273	227	194	209	286	295	230	226	199	297	308	226	215	184	299	315	215	132	83	229	245	123	9
	12	209	194	259	285	259	194	178	270	308	270	178	157	273	321	273	157	132	268	324	268	132	9	182	253	182	9
Feb. 21	7/5	22	64	60	34	7	7	80	75	37	7	7	94	88	40	6	6	106	98	42	6	6	109	109?	38	6	6
	8/4	95	168	172	127	60	14	194	199	136	41	14	214	220	140	21	14	228	235	141	12	12	218	228	108	7	7
	9/3	161	225	243	206	136	75	244	269	217	118	75	256	287	222	97	59	261	297	209	74	14	209	256	158	9	9
	10/2	212	254	286	266	206	141	261	306	278	193	141	261	318	283	175	130	253	320	279	152	15	160	246	193	32	9
	11/1	245	257	302	304	260	197	251	315	317	255	197	237	320	321	243	188	218	313	315	223	91	90	211	214	98	9
	12	255	237	293	316	293	237	217	297	330	297	217	192	292	334	292	192	160	279	328	279	160	10	160	222	160	10
Mar. 21	7/5	54	117	104	60	13	13	141	123	60	12	12	161	138	59	11	11	175	149	56	11	11	185	151	32	7	7
	8/4	129	198	198	146	79	36	230	221	147	52	17	249	237	144	24	16	260	247	137	15	15	235	217	78	9	9
	9/3	194	264	264	222	155	104	275	284	224	130	87	284	295	220	102	61	285	298	209	70	17	206	232	119	10	10
	10/2	245	287	305	280	225	171	290	312	283	205	162	284	323	278	179	133	275	317	265	148	35	156	217	150	11	11
	11/1	277	287	321	317	278	227	279	319	321	266	226	262	322	315	246	185	238	308	300	218	111	89	180	170	67	11
	12	287	267	311	329	311	267	246	308	333	308	246	217	295	327	295	217	181	274	312	274	181	14	128	177	128	11
Apr. 21	6/6	14	37	27	9	6	6	46	32	6	6	6	54	36	6	6	6	61	40	5	5	5	68	42	3	3	3
	7/5	86	153	130	78	27	19	178	145	71	18	23	197	155	62	17	17	210	161	51	16	16	205	142	10	12	10
	8/4	158	211	211	156	87	66	250	225	148	63	86	265	234	136	29	22	272	235	120	20	20	229	151	35	15	12
	9/3	220	277	270	225	168	133	290	281	217	137	154	295	284	203	102	38	285	278	183	64	22	206	188	68	17	13
	10/2	267	304	308	279	233	196	294	312	272	206	214	295	307	256	173	58	282	294	234	136	58	156	171	95	30	14
	11/1	297	310	317	313	280	248	290	319	302	263	263	275	306	290	235	174	247	285	265	201	129	89	136	112	58	17
	12	307	286	311	325	311	286	263	302	318	302	263	232	281	301	281	232	195	252	276	252	195	14	88	118	88	14
May 21	6/6	36	76	54	21	14	14	91	60	13	13	13	103	65	13	13	13	113	68	12	12	12	119	62	7	9	7
	7/5	107	172	142	88	41	29	194	152	75	24	29	211	158	60	22	26	222	159	44	21	21	204	124	13	14	13
	8/4	174	209	214	154	107	87	258	223	139	71	105	270	225	127	33	26	274	221	105	21	24	218	151	33	16	15
	9/3	233	284	264	223	174	150	295	267	209	139	168	298	270	188	101	44	290	259	163	59	26	195	138	56	17	16
	10/2	277	308	304	273	234	209	307	290	259	203	224	298	291	237	167	76	281	272	208	126	74	148	105	72	17	17
	11/1	305	310	317	305	280	258	290	292	290	256	267	278	290	268	224	185	249	263	237	186	140	87	59	77	17	17
	12	315	293	309	315	304	293	270	285	292	285	270	238	267	278	267	238	201	234	247	234	201	17	48	60	48	19
June 21	6/6	45	87	62	26	17	17	103	68	16	16	16	116	72	15	15	15	125	72	14	14	14	128	62	9	9	9
	7/5	115	177	145	91	46	38	198	152	76	26	26	213	156	59	25	23	222	155	41	23	23	200	114	14	14	14
	8/4	180	241	213	159	116	95	258	219	143	73	112	268	219	122	52	27	271	213	99	27	27	211	138	16	18	16
	9/3	236	284	266	221	175	156	293	267	201	139	177	294	261	181	100	58	287	248	153	59	29	189	140	41	18	18
	10/2	279	307	299	269	233	213	306	282	251	201	227	294	276	237	163	80	278	261	197	121	80	144	123	56	18	18
	11/1	306	310	312	299	277	260	292	281	281	257	270	276	281	257	218	185	248	253	224	179	143	85	92	72	19	19
	12	315	293	304	309	304	293	270	285	292	285	270	239	259	267	259	239	201	224	234	224	201	19	48	60	48	19

178

BTU/FT² Hourly Total Insolation, $I_{DH} + I_d$, On Surfaces Tilted With Respect To The Horizontal

Date	Solar Time AM	Solar Time PM	0° S	22° E/W	22° SE/SW	22° S	22° SW/SE	22° W/E	32° E/W	32° SE/SW	32° S	32° SW/SE	32° W/E	42° E/W	42° SE/SW	42° S	42° SW/SE	42° W/E	52° E/W	52° SE/SW	52° S	52° SW/SE	52° W/E	90° E/W	90° SE/SW	90° S	90° SW/SE	90° W/E
July 21	6	6	37	74	54	22	15	15	88	59	14	14	14	100	63	13	13	13	109	66	12	12	12	114	59	8	8	8
	7	5	107	169	140	87	43	32	190	149	75	25	25	206	154	60	24	24	216	154	44	27	22	197	119	14	14	14
	8	4	174	235	211	158	107	89	253	218	143	72	46	264	220	125	35	29	267	215	104	60	27	212	146	16	16	16
	9	3	231	280	264	220	173	150	290	268	205	139	107	291	285	185	101	60	285	253	159	125	29	190	149	31	18	18
	10	2	274	303	299	269	208	208	294	302	254	201	168	274	284	232	165	123	276	267	204	183	76	146	133	54	19	19
	11	1	302	306	312	300	276	256	303	287	285	252	223	298	262	262	221	194	246	258	232	229	140	86	102	69	130	19
	12		311	290	304	310	290	290	267	287	296	287	267	274	262	273	236	236	199	229	242	229	199	19	58	74	58	19
Aug. 21	6	6	14	35	26	9	7	7	43	30	7	7	7	50	34	6	6	6	55	36	5	5	6	62	37	4	4	4
	7	5	85	147	125	77	30	22	169	138	69	21	27	186	147	32	20	20	198	152	50	19	20	191	132	12	12	12
	8	4	156	221	204	152	97	69	242	217	144	65	89	255	224	132	25	25	261	225	116	24	25	217	171	33	15	15
	9	3	216	270	263	220	166	133	282	272	212	136	89	286	274	197	102	43	283	268	178	65	43	198	179	65	16	17
	10	2	262	295	300	272	228	194	296	306	264	202	154	289	298	249	171	109	274	276	226	134	62	151	164	91	17	17
	11	1	292	299	314	305	272	245	288	310	299	247	212	289	274	281	230	173	247	276	257	197	130	88	131	107	30	17
	12		302	281	306	317	306	281	258	294	309	294	258	229	274	292	274	229	192	246	268	246	192	17	85	113	85	17
Sept. 21	7	5	51	107	95	56	14	14	128	112	56	14	14	145	125	55	13	13	158	134	52	12	12	165	135	30	7	7
	8	4	124	194	188	140	78	38	217	209	141	53	20	234	224	138	27	19	245	232	131	18	19	219	203	75	11	11
	9	3	188	247	253	213	151	103	263	271	215	127	59	271	282	268	100	84	272	284	201	144	80	205	208	114	13	13
	10	2	237	274	294	270	218	167	279	306	273	199	127	276	310	303	174	152	264	304	289	212	89	157	174	144	66	14
	11	1	268	278	309	306	269	221	270	314	309	257	189	254	310	315	238	210	230	296	306	264	110	90	125	164	125	14
	12		278	259	301	318	301	259	238	297	321	297	238	210	285	315	285	210	177	264	318	264	177	14	171	171	171	14
Oct. 21	7	5	19	53	50	29	7	7	66	63	32	7	7	78	73	34	6	6	87	81	36	6	6	97	90	32	4	4
	8	4	90	157	161	120	58	16	181	187	128	40	15	200	207	133	22	15	213	221	134	14	14	204	214	104	8	8
	9	3	155	215	232	198	132	73	233	257	208	115	32	245	275	273	95	52	249	285	212	149	89	205	245	153	34	10
	10	2	204	244	275	257	199	136	251	295	269	188	99	251	307	303	171	127	243	309	270	217	89	155	238	188	34	11
	11	1	247	248	292	294	252	191	242	305	307	247	161	229	308	316	236	211	210	303	306	271	156	88	206	209	97	11
	12		229	284	306	320	288	267	211	288	320	288	211	186	283	324	285	186	156	271	318	271	156	11	156	217	156	11
Nov. 21	7	5	0	1	1	1	0	0	1	1	1	0	0	1	1	1	0	0	1	1	1	0	0	2	2	1	0	0
	8	4	55	112	122	91	37	12	133	147	104	27	11	150	168	113	17	11	163	184	119	10	10	166	194	111	6	6
	9	3	118	175	200	173	108	44	194	229	190	99	15	206	252	202	86	14	213	267	208	72	13	183	251	176	60	8
	10	2	166	206	246	225	166	104	214	271	252	169	130	217	289	265	160	135	211	297	270	146	67	145	252	217	121	9
	11	1	207	211	256	270	225	193	207	282	291	228	177	198	293	303	223	100	181	294	307	213	131	83	225	241	179	10
	12			193		282	256		177	267	304	267	156	156	270	316	270	156	131	264	320	264	131	10	179	249	179	10
Dec. 21	8	4	41	91	103	77	29	10	110	127	90	22	9	126	147	101	14	9	138	162	108	8	8	145	176	107	5	5
	9	3	102	158	186	161	97	31	176	216	180	91	14	189	240	195	82	13	197	256	204	70	12	174	249	183	13	13
	10	2	150	189	232	221	162	90	198	259	244	160	58	202	279	259	154	25	194	290	267	143	58	140	256	226	69	8
	11	1	180	194	250	258	213	140	192	271	282	218	116	184	294	298	217	88	170	289	305	209	120	80	231	251	130	9
	12		190	177	243	271	243	177	162	256	295	256	162	143	262	311	262	143	120	260	318	260	120	9	186	259	186	9

aBased on data in Table 1, p. 387, 1972 ASHRAE Handbook of Fundamentals; 0% ground reflectance, 1.0 clearness factor.

179

Table A.3. Btu/ft² Insolation on Inclined Surfaces at 40° North Latitude[a]

Btu/Ft² Hourly Total Insolation, $I_{D\theta B} + I_d$, On Surfaces Tilted With Respect To The Horizontal

Date	Solar Time AM	PM	I_D	30° E/W	30° SE/SW	30° S	30° SW/SE	30° W/E	40° E/W	40° SE/SW	40° S	40° SW/SE	40° W/E	50° E/W	50° SE/SW	50° S	50° SW/SE	50° W/E	60° E/W	60° SE/SW	60° S	60° SW/SE	60° W/E	90° E/W	90° SE/SW	90° S	90° SW/SE	90° W/E
Jan 21	8	4	28	82	94	65	12	8	96	111	74	7	7	108	125	81	7	7	116	135	85	6	6	119	142	84	4	4
	9	3	83	152	187	155	75	13	167	212	171	68	12	177	231	182	59	11	179	243	187	48	10	166	235	171	11	7
	10	2	127	175	232	218	141	46	182	255	237	138	16	183	271	249	131	13	179	278	254	120	12	137	251	223	69	8
	11	1	154	170	246	257	196	100	165	263	277	199	75	156	273	290	197	48	142	274	293	188	20	79	231	253	132	8
	12		164	143	231	270	233	143	127	243	291	243	127	108	246	303	246	108	86	242	306	242	86	9	188	263	188	9
Feb 21	7	5	10	42	39	14	4	4	50	47	21	4	4	57	54	26	3	3	63	59	24	3	3	68	63	22	2	2
	8	4	73	159	167	114	32	13	180	190	122	16	12	196	207	126	11	11	205	218	127	10	11	198	213	107	7	7
	9	3	132	211	239	195	104	20	225	262	205	89	14	234	277	209	71	13	235	284	208	51	13	199	255	167	8	8
	10	2	178	228	277	256	175	82	232	296	267	163	45	239	305	271	148	13	235	305	267	127	13	154	254	210	48	9
	11	1	206	219	288	290	233	141	210	306	306	228	110	195	301	310	217	76	175	274	304	199	40	87	225	236	115	9
	12		216	189	272	306	272	189	168	275	319	275	168	142	270	323	270	142	113	257	317	257	113	9	176	245	176	9
Mar 21	7	5	46	123	109	55	11	11	143	125	55	11	11	158	137	54	10	10	169	145	51	9	9	171	143	35	6	6
	8	4	114	208	205	140	51	17	228	224	141	27	18	242	237	138	15	15	249	243	131	13	13	228	219	89	9	9
	9	3	173	251	267	215	126	51	263	283	217	102	18	267	292	213	76	16	264	292	202	48	15	209	242	138	10	10
	10	2	218	265	301	273	196	116	265	313	276	177	74	258	314	271	153	30	243	307	258	125	16	159	233	176	22	11
	11	1	247	255	310	310	254	176	242	313	313	242	141	223	308	307	222	102	197	293	293	196	61	90	200	200	89	11
	12		257	224	294	322	294	224	199	289	326	289	199	169	276	320	276	169	134	255	305	255	134	11	150	208	150	11
Apr 21	6	6	20	62	44	11	8	8	73	50	8	8	8	82	55	7	7	7	88	57	7	7	7	92	57	4	4	4
	7	5	87	174	146	77	19	13	194	158	70	18	18	209	166	61	16	16	217	169	50	15	15	205	149	12	10	10
	8	4	152	240	223	150	72	20	256	235	145	40	22	266	240	133	20	20	267	238	117	18	18	226	192	50	12	12
	9	3	207	276	277	221	149	26	284	285	213	132	40	283	284	199	78	22	275	275	177	43	20	203	204	93	13	13
	10	2	250	289	308	270	209	86	285	305	267	183	125	273	305	252	151	29	253	288	229	115	22	154	225	126	43	14
	11	1	277	279	315	301	263	169	263	305	301	242	184	240	285	285	215	127	210	272	260	182	82	88	160	147	104	14
	12		287	250	299	320	299	250	223	286	313	286	223	189	265	296	265	189	150	236	271	236	150	14	113	154	113	14
May 21	5	7	2	1	0	0	0	0	1	0	0	0	0	1	0	0	0	0	1	0	0	0	0	1	0	0	0	0
	6	6	49	111	78	25	16	16	126	84	15	15	15	138	87	14	14	14	146	88	13	13	13	144	77	9	9	9
	7	5	114	198	162	89	52	24	216	169	76	50	23	228	172	60	25	21	233	170	44	23	20	209	136	13	13	13
	8	4	175	255	229	158	110	85	268	234	144	117	27	273	233	125	80	25	271	226	104	40	24	218	166	50	15	15
	9	3	227	288	277	221	169	152	292	279	206	183	64	288	272	186	147	27	276	258	160	107	29	194	173	89	16	16
	10	2	267	299	306	270	226	213	293	301	255	238	125	278	281	233	206	78	255	265	205	168	95	147	160	126	33	17
	11	1	293	290	312	301	264	232	275	307	287	279	184	247	288	264	252	147	214	270	234	219	159	86	136	147	100	17
	12		301	263	298	312	298	263	235	279	297	279	235	200	252	274	252	200	159	219	243	219	159	17	86	114	86	17
June 21	5	7	4	14	7	3	3	3	16	8	3	3	3	18	8	2	2	2	19	9	2	2	2	21	8	1	1	1
	6	6	60	125	87	30	19	19	140	92	18	18	18	151	94	17	17	17	158	93	16	16	16	152	77	10	10	10
	7	5	123	205	165	92	62	27	221	170	77	54	26	231	170	59	24	24	235	188	41	23	23	206	153	14	14	14
	8	4	182	258	228	159	119	90	269	231	142	119	73	277	228	121	80	29	259	248	97	40	29	212	159	16	16	16
	9	3	233	289	275	219	175	154	292	273	202	181	132	287	264	179	144	86	259	256	151	107	103	188	146	47	18	18
	10	2	272	300	302	262	226	213	292	295	248	235	189	277	279	224	201	147	234	265	194	162	147	143	117	92	24	18
	11	1	296	292	308	296	262	238	273	294	278	274	235	247	272	253	245	201	214	243	221	210	162	84	75	98	75	19
	12		304	266	294	306	294	266	238	274	289	274	238	202	245	263	245	202	162	210	230	210	162	19	75	98	75	19

Date	AM	PM	0°	30° E/W	30° SE/SW	30° S/S	30° SW/SE	30° W/E	40° E/W	40° SE/SW	40° S/S	40° SW/SE	40° W/E	50° E/W	50° SE/SW	50° S/S	50° SW/SE	50° W/E	60° E/W	60° SE/SW	60° S/S	60° SW/SE	60° W/E	90° E/W	90° SE/SW	90° S/S	90° SW/SE	90° W/E
July 21	5	7	0	1	1	0	0	0	2	1	0	0	0	2	1	0	0	0	2	1	0	0	0	2	1	0	0	0
	6	6	57	109	77	26	17	17	123	82	17	17	17	134	85	15	15	15	141	86	14	14	14	138	74	9	9	9
	7	5	114	195	159	89	26	26	212	165	75	21	25	222	167	60	23	23	227	165	44	21	21	202	130	14	14	14
	8	4	174	251	215	157	86	51	263	230	142	52	29	268	228	124	27	27	265	220	102	25	25	212	160	24	16	16
	9	3	225	283	273	218	151	112	287	273	203	117	66	283	266	182	80	29	270	251	157	41	26	189	168	58	18	18
	10	2	265	295	300	266	211	169	288	295	251	181	126	273	282	229	145	80	251	260	200	106	32	144	126	86	18	18
	11	1	290	287	307	296	260	220	269	295	281	234	183	244	275	258	203	142	212	247	228	166	97	85	104	104	32	18
	12	12	298	261	293	307	293	261	233	274	292	274	233	198	248	269	248	198	158	215	238	215	158	19	84	111	84	19
Aug 21	6	6	21	58	42	12	9	9	68	47	9	9	9	76	51	8	8	8	81	53	7	8	8	84	51	5	5	5
	7	5	87	167	140	76	22	22	185	151	69	21	21	198	157	60	19	19	205	159	49	19	19	192	139	12	12	12
	8	4	150	232	215	150	74	32	247	226	141	43	26	255	229	129	24	24	256	227	113	24	24	215	181	50	14	14
	9	3	205	269	268	216	142	90	276	275	207	112	45	275	274	193	80	26	265	265	173	45	26	195	194	89	16	16
	10	2	246	282	299	267	205	149	278	300	259	179	107	266	292	244	148	63	246	276	221	114	45	149	184	120	17	17
	11	1	273	269	307	300	257	203	258	301	292	234	168	235	286	276	210	128	206	263	252	177	84	87	154	140	54	17
	12	12	282	246	292	311	292	246	220	279	303	279	220	187	258	287	258	187	149	229	262	229	149	17	109	147	109	17
Sept 21	7	5	43	111	98	51	13	13	128	112	51	12	12	141	122	49	11	11	150	129	47	10	10	151	127	32	3	3
	8	4	109	196	193	133	52	52	214	210	134	29	29	227	222	131	17	17	232	227	124	16	16	210	204	84	11	11
	9	3	167	239	254	206	123	72	250	270	208	100	52	254	277	203	76	31	251	277	193	49	19	198	228	132	19	15
	10	2	211	254	289	262	190	115	255	299	265	172	100	247	300	260	149	76	233	293	247	122	49	153	222	168	50	19
	11	1	239	246	298	298	246	172	234	301	301	234	162	215	295	295	215	131	191	281	281	190	84	88	192	192	114	41
	12	12	249	217	283	310	283	217	194	278	313	278	194	165	265	307	265	165	131	245	292	245	131	13	145	200	145	13
Oct 21	7	5	7	30	28	28	14	14	36	34	15	14	14	41	38	17	3	3	44	42	17	3	3	48	45	16	3	3
	8	4	68	147	154	106	31	21	165	175	113	21	21	179	191	117	12	13	188	201	118	11	11	181	196	100	11	11
	9	3	126	199	227	185	100	31	213	248	195	86	44	221	263	200	69	17	222	269	198	50	15	188	244	160	15	15
	10	2	170	218	266	245	168	44	222	283	258	158	80	219	290	261	143	44	210	293	257	124	41	148	244	203	50	17
	11	1	199	211	277	283	225	100	202	288	295	220	107	188	290	299	210	107	169	284	294	193	75	85	217	229	119	41
	12	12	208	182	262	295	262	182	162	265	308	265	162	138	261	312	261	138	110	248	306	248	110	11	171	238	171	11
Nov 21	8	4	28	80	91	63	13	13	94	108	72	8	8	105	121	78	7	7	112	131	82	6	6	115	137	81	4	4
	9	3	82	150	183	152	74	17	164	208	167	67	14	174	226	178	58	14	179	237	183	48	11	162	229	167	11	11
	10	2	126	173	229	215	140	47	180	251	233	137	47	185	269	245	129	49	176	273	249	119	48	135	246	219	68	13
	11	1	153	169	243	254	194	100	164	260	273	197	108	155	269	285	194	90	141	270	288	185	75	78	227	248	129	21
	12	12	163	142	230	267	230	142	127	240	287	240	127	108	243	298	243	108	86	238	301	238	86	9	185	258	185	9
Dec 21	8	4	14	47	56	39	14	14	56	67	45	4	4	63	76	50	4	4	69	84	54	4	4	73	90	56	3	3
	9	3	65	127	162	135	62	17	141	186	152	58	11	152	205	164	52	10	157	218	171	45	9	147	216	163	17	6
	10	2	107	154	212	200	127	34	161	235	221	126	34	168	252	235	122	12	161	261	242	114	11	127	243	221	73	7
	11	1	134	150	227	239	180	84	147	246	262	186	84	140	257	276	186	38	127	261	283	180	14	74	227	252	134	8
	12	12	143	125	215	253	215	125	112	227	275	227	112	95	233	290	233	95	76	232	296	232	76	8	188	263	188	8

aBased on data in Table 1, p. 387, *1972 ASHRAE Handbook of Fundamentals*; 0% ground reflectance, 1.0 clearness factor.

Table A.4. Btu/ft² Insolation on Inclined Surfaces at 48° North Latitude[a]

BTU/ft² Hourly Total Insolation, $I_{DH} + I_d$, On Surfaces Tilted With Respect To The Horizontal

Date	Solar Time AM	Solar Time PM	0°	38° E/W	38° SE/SW	38° S	38° SW/SE	38° W/E	48° E/W	48° SE/SW	48° S	48° SW/SE	48° W/E	58° E/W	58° SE/SW	58° S	58° SW/SE	58° W/E	68° E/W	68° SE/SW	68° S	68° SW/SE	68° W/E	90° E/W	90° SE/SW	90° S	90° SW/SE	90° W/E
Jan. 21	8	4	4	22	26	17		2	26	30	19	2	2	28	34	21	1	2	30	36	22	1	1	31	37	22	1	1
	9	3	46	113	149	120	42	10	124	167	132	38	12	131	181	140	34	8	135	189	145	28	7	128	187	139	13	5
	10	2	83	136	216	190	105	12	142	222	206	104	27	143	234	216	100	11	140	240	220	93	10	119	227	206	68	7
	11	1	107	125	216	231	161	47	121	231	249	165	80	114	240	260	164	12	121	241	263	158	10	71	219	243	129	8
		12	115	93	200	245	200	93	80	210	264	210	80	65	213	275	213	65	48	210	278	210	48	8	183	255	183	8
Feb. 21	7	5	4	8	8	3	1	1	9	9	4	1	1	10	10	4	1	1	11	11	4			11	11	4		
	8	4	49	138	149	95	13	10	154	166	102	9	10	164	179	105	9	13	170	186	106	8	10	183	183	96	8	6
	9	3	100	188	226	178	73	13	199	245	187	61	13	205	257	191	47	14	204	261	190	32	11	204	244	167	59	8
	10	2	139	196	262	240	143	27	198	277	251	134	14	194	285	257	121	14	184	284	251	104	12	145	252	217	125	9
	11	1	165	177	268	278	203	86	168	277	290	200	58	136	279	294	190	58	136	272	288	175	70	83	230	247	185	9
		12	173	138	247	291	247	138	119	249	304	249	119	96	245	307	245	96	70	233	301	233	70	9	185	258	185	9
Mar. 21	7	5	37	121	108	49	10	10	136	120	49	9	10	147	128	47	8	9	154	133	45	7	7	153	131	35	5	5
	8	4	96	203	204	131	26	15	219	220	132	14	18	228	230	129	8	17	230	232	122	12	14	213	215	96	10	10
	9	3	147	236	263	205	97	17	244	277	207	76	20	245	282	203	53	20	238	280	193	29	16	201	242	152	39	12
	10	2	187	238	292	263	167	61	235	301	266	150	22	226	301	261	129	22	210	292	248	103	19	154	242	195	106	13
	11	1	212	216	295	300	228	122	202	297	303	219	88	182	290	297	199	88	157	275	283	175	89	87	217	223	167	17
		12	220	176	276	312	276	176	151	267	315	267	151	122	254	309	254	122	89	234	294	234	89	11	167	232	167	11
Apr. 21	6	6	27	87	62	13	9	9	98	67	9	9	9	106	71	8	8	9	111	73	7	7	7	111	70	5	5	5
	7	5	85	189	158	76	18	18	204	168	69	17	17	214	173	59	15	17	218	172	48	14	14	204	155	21	12	12
	8	4	142	243	230	149	38	24	270	240	141	20	22	258	240	129	18	22	255	235	113	16	18	221	200	69	13	13
	9	3	191	267	278	217	128	66	259	283	260	158	27	244	293	245	95	27	223	268	223	95	19	199	217	152	15	15
	10	2	228	267	303	268	191	123	227	302	281	219	78	201	281	278	128	78	176	257	254	162	36	180	209	177	78	17
	11	1	252	246	303	301	240	183	178	297	292	263	138	144	247	289	247	144	135	218	264	213	106	135	180	185	110	17
		12	260	218	282	313	282	218	194	268	305	268	194	178	247		247	178	135	218		218	135	14	135	135	14	14
May 21	5	7	9	30	16	4	4	9	34	17	4	4	4	37	18	4	4	4	39	18	4	3	3	40	17	2	2	2
	6	6	61	144	100	27	18	18	157	104	16	16	16	166	106	15	15	16	170	104	13	13	13	162	91	10	10	10
	7	5	118	218	177	89	24	24	231	182	75	22	22	237	181	60	20	22	236	176	43	18	18	212	146	13	13	13
	8	4	171	262	239	156	63	27	269	241	142	29	27	250	237	123	23	27	261	225	101	21	15	217	179	45	15	15
	9	3	217	282	281	217	128	66	281	279	202	160	78	250	280	229	128	78	224	257	200	88	19	192	190	120	17	17
	10	2	252	282	303	265	191	123	230	296	281	219	78	211	268	258	187	78	176	239	228	150	25	185	153	141	57	17
	11	1	274	261	303	296	244	183	194	290	281	263	138	157	269	278	236	157	116	202	238	202	50	17	135	149	110	17
		12	281	230	283	306	283	230	225	263	292	263	194		269		269	157	135	202	238	202	116	161	135	149	110	
June 21	5	7	21	60	32	9	21	9	66	34	9	19	9	71	34	8	18	9	74	33	7	16	7	74	29	5	12	5
	6	6	74	158	110	33	26	29	171	112	19	19	25	179	112	15	18	25	182	109	16	16	20	170	91	5	15	15
	7	5	129	225	181	157	69	29	236	183	77	34	27	241	180	59	25	30	238	172	39	24	23	209	138	5	16	16
	8	4	181	266	239	216	132	77	271	239	140	97	27	269	232	119	60	27	259	217	95	24	25	212	167	4	17	17
	9	3	225	285	279	216	132	216	282	275	198	132	41	271	263	220	124	41	253	243	189	85	25	186	177	120	48	17
	10	2	259	284	300	262	192	132	242	284	244	183	15	212	261	248	183	15	177	230	216	150	56	142	167	141	48	17
	11	1	280	265	300	291	243	185	198	259	273	219	138	161	258		230	161	119	194	225	194	119	18	99	149	99	18
		12	287	230	280	301	280	230	198	259	283	259	194		258		258		119	194		194		18				18

BTU/FT² Hourly Total Insolation, $I_{D\theta S} + I_d$, On Surfaces Tilted With Respect To The Horizontal

Date	AM	PM	0°	38° E	38° SE	38° S	38° SW	38° W	48° E	48° SE	48° S	48° SW	48° W	58° E	58° SE	58° S	58° SW	58° W	68° E	68° SE	68° S	68° SW	68° W	90° E	90° SE	90° S	90° SW	90° W
July 21	5	7	10	32	18	5	5	5	36	19	5	5	5	40	19	4	4	4	42	19	4	4	4	42	18	3	3	3
	6	6	62	141	99	89	19	19	153	102	18	18	18	162	103	16	16	16	165	101	15	15	15	156	88	11	11	11
	7	5	118	214	174	154	26	26	226	178	75	24	24	231	177	59	22	22	230	171	42	19	19	205	141	14	14	14
	8	4	171	258	234	214	65	29	264	236	140	31	27	263	231	121	25	25	254	219	99	23	23	211	173	43	16	16
	9	3	215	278	276	261	128	69	276	274	199	95	29	266	266	178	60	27	249	246	153	56	25	187	184	83	17	17
	10	2	250	277	297	291	189	124	266	284	246	159	81	246	274	224	125	33	220	251	183	125	30	143	175	116	18	18
	11	1	272	224	298	301	241	177	237	284	276	215	139	208	263	253	184	97	174	234	214	184	85	84	148	137	56	18
	12		279	205	278	304	278	224	193	259	286	259	193	157	232	263	232	157	116	198	223	198	116	19	107	144	107	19
Aug. 21	6	6	28	82	59	14	11	11	92	64	10	10	10	99	67	4	9	9	103	68	8	8	8	102	64	6	6	6
	7	5	85	180	152	75	21	21	194	160	67	19	19	203	163	58	18	18	206	162	47	16	16	191	144	20	12	12
	8	4	141	235	222	145	51	24	245	229	137	24	24	248	230	125	22	22	244	224	115	19	19	210	189	65	14	14
	9	3	189	260	269	210	116	44	262	273	201	89	26	256	269	187	58	24	243	257	181	94	27	191	206	110	15	15
	10	2	225	261	295	260	181	101	253	292	252	156	59	236	282	237	127	46	215	264	233	168	30	146	200	146	16	16
	11	1	248	241	295	293	235	157	223	288	285	215	120	198	272	268	189	116	167	248	265	224	81	85	173	169	75	17
	12		256	205	275	304	275	205	177	261	296	261	177	157	239	279	239	182	106	211	276	211	106	17	130	177	130	17
Sept. 21	7	5	35	107	95	44	11	11	120	105	44	10	10	129	113	43	9	9	134	117	40	8	8	133	114	31	6	6
	8	4	92	189	190	124	28	18	203	205	124	17	17	211	213	121	15	15	213	215	115	14	14	196	198	90	10	10
	9	3	142	224	249	196	94	21	231	261	197	75	19	231	266	193	53	19	225	263	183	30	17	189	230	143	12	12
	10	2	181	228	279	251	162	62	225	287	254	145	25	216	286	248	125	25	200	277	236	118	18	147	230	185	40	13
	11	1	205	208	282	287	220	120	195	284	289	208	88	176	272	284	191	54	151	262	269	168	54	85	204	212	103	13
	12		213	171	261	299	261	171	147	256	302	256	147	119	244	296	244	119	88	224	281	224	13	13	160	221	160	13
Oct. 21	7	5	3	3	2	1	0	0	3	3	1	0	0	3	3	1	0	0	4	3	1	0	0	4	4	1	0	0
	8	4	44	123	132	86	11	11	136	148	91	10	10	146	159	95	9	9	151	165	94	8	8	147	163	87	6	6
	9	3	94	175	211	167	70	15	186	229	176	59	14	191	240	180	46	14	190	235	178	101	12	170	228	157	8	8
	10	2	133	185	248	228	137	27	187	263	239	128	27	184	269	241	116	29	175	260	239	118	14	138	240	207	58	10
	11	1	157	169	255	266	195	83	160	265	277	191	57	147	264	281	182	93	130	260	276	168	93	80	220	237	121	10
	12		166	133	236	279	236	133	114	239	291	239	114	93	235	294	235	93	68	224	288	224	68	10	178	247	178	10
Nov. 21	8	4	5	22	26	17	2	2	26	30	19	2	2	28	33	21	2	2	30	36	22	2	2	31	37	22	1	1
	9	3	46	111	146	117	42	10	123	163	120	38	9	128	176	137	33	9	131	184	141	27	8	125	181	135	13	7
	10	2	83	134	192	186	104	28	139	217	202	102	28	140	229	212	98	11	137	235	215	91	10	117	222	201	67	8
	11	1	107	124	213	227	159	48	120	228	245	162	80	113	255	255	161	65	103	236	258	155	48	70	214	238	127	8
	12		115	92	198	241	198	92	80	207	259	207	80	65	210	270	210	65	48	207	272	207	48	8	179	250	179	8
Dec. 21	9	3	27	78	108	87	28	7	87	122	98	27	7	93	134	105	25	6	97	141	110	22	5	95	142	109	14	4
	10	2	63	111	173	164	88	17	117	192	180	89	17	119	205	192	87	11	118	212	197	83	8	104	205	190	66	6
	11	1	86	105	192	207	142	34	103	208	226	148	65	98	218	239	149	53	90	221	244	146	40	64	206	231	125	7
	12		94	75	179	222	179	74	65	190	241	190	65	53	195	254	195	53	40	195	260	195	40	7	175	244	175	7

[a] Based on data in Table 1, p. 387, *1972 ASHRAE Handbook of Fundamentals;* 0% ground reflectance, 1.0 clearness factor.

183

Table A.5. Btu/ft² Insolation on Inclined Surfaces at 56° North Latitude[a]

BTU/ft² Hourly Total Insolation, $I_{DS} + I_d$, On Surfaces Tilted With Respect To The Horizontal

Orientation columns within each tilt angle are given as E/W, SE/SW, S/S, SW/SE, W/E (AM orientation / PM mirror).

Jan 21

AM	PM	0°	46° E	46° SE	46° S	46° SW	46° W	56° E	56° SE	56° S	56° SW	56° W	66° E	66° SE	66° S	66° SW	66° W	76° E	76° SE	76° S	76° SW	76° W	90° E	90° SE	90° S	90° SW	90° W
9	3	11	46	64	50	12	4	50	71	55	11	4	53	77	59	10	3	55	79	60	9	3	54	80	60	7	2
10	2	39	86	144	135	63	8	90	158	146	64	9	92	166	154	62	7	91	169	156	60	6	85	166	153	53	5
11	1	58	79	167	183	116	10	77	179	197	120	9	73	186	206	121	8	67	187	208	118	7	56	180	201	108	6
12		65	47	154	198	154	47	39	162	214	162	39	30	166	222	166	30	21	165	225	165	21	6	155	217	155	6

Feb 21

AM	PM	0°	46° E	46° SE	46° S	46° SW	46° W	56° E	56° SE	56° S	56° SW	56° W	66° E	66° SE	66° S	66° SW	66° W	76° E	76° SE	76° S	76° SW	76° W	90° E	90° SE	90° S	90° SW	90° W
8	4	25	97	107	65	7	7	106	118	69	6	6	113	125	72	5	5	115	129	72	5	5	114	127	69	4	4
9	3	65	155	197	151	45	11	163	211	159	37	10	167	220	162	27	9	165	222	161	17	8	156	214	151	6	6
10	2	98	159	235	215	110	13	160	248	225	104	11	157	257	228	95	11	149	263	224	83	9	131	237	208	63	7
11	1	119	134	239	254	170	36	126	248	265	168	14	115	257	268	161	11	101	263	263	150	10	76	222	243	126	8
12		126	90	216	268	216	90	74	219	279	219	74	56	216	282	216	56	37	206	276	206	37	8	183	255	183	8

Mar 21

AM	PM	0°	46° E	46° SE	46° S	46° SW	46° W	56° E	56° SE	56° S	56° SW	56° W	66° E	66° SE	66° S	66° SW	66° W	76° E	76° SE	76° S	76° SW	76° W	90° E	90° SE	90° S	90° SW	90° W
7	5	28	109	97	40	8	8	120	83	40	9	9	127	112	39	6	6	130	114	37	6	6	128	111	32	5	5
8	4	75	189	195	119	13	15	200	173	120	12	15	205	213	117	11	13	205	213	111	7	11	194	203	107	8	9
9	3	118	214	252	192	69	15	219	237	193	52	15	217	265	189	34	14	209	260	180	14	12	188	241	158	9	10
10	2	151	206	277	249	138	16	202	276	251	124	15	192	282	246	106	15	176	272	234	85	12	146	244	205	52	10
11	1	172	175	274	285	200	70	161	283	288	190	40	143	269	282	175	40	120	254	268	154	51	83	221	236	118	17
12		179	128	248	297	248	128	105	243	300	243	105	79	231	294	231	79	51	213	280	213	51	10	177	246	177	10

Apr 21

AM	PM	0°	46° E	46° SE	46° S	46° SW	46° W	56° E	56° SE	56° S	56° SW	56° W	66° E	66° SE	66° S	66° SW	66° W	76° E	76° SE	76° S	76° SW	76° W	90° E	90° SE	90° S	90° SW	90° W
5	7	0	0	0	0	0	0	0	0	0	0	0	0	0	0	0	0	0	0	0	0	0	0	0	0	0	0
6	6	32	110	78	14	10	10	120	83	66	9	9	126	85	57	8	8	129	85	46	7	7	126	81	29	6	6
7	5	81	196	166	74	17	17	207	173	135	15	15	213	174	123	14	14	211	171	108	12	12	200	158	82	10	11
8	4	129	233	232	143	25	22	245	237	193	18	21	245	236	186	16	16	237	227	167	14	14	215	205	133	12	13
9	3	169	239	274	209	97	27	246	276	251	65	21	245	270	236	58	18	228	257	214	77	16	193	225	174	13	13
10	2	201	208	293	259	157	48	228	290	284	134	48	225	279	268	107	33	188	260	245	143	17	146	221	200	85	14
11	1	220	162	288	292	216	106	188	280	295	197	106	182	264	268	173	100	132	240	245	163	130	84	195	200	154	96
12		227	127	261	303	261	162	133	248	295	248	183	130	227	279	227	133	65	200	255	200	200	14	152	209	152	152

May 21

AM	PM	0°	46° E	46° SE	46° S	46° SW	46° W	56° E	56° SE	56° S	56° SW	56° W	66° E	66° SE	66° S	66° SW	66° W	76° E	76° SE	76° S	76° SW	76° W	90° E	90° SE	90° S	90° SW	90° W
4	8	0	0	0	0	0	0	0	0	0	0	0	0	0	0	0	0	0	0	0	0	0	0	0	0	0	0
5	7	25	80	43	0	0	10	86	44	9	9	9	90	44	8	8	8	92	43	7	7	7	90	39	6	6	6
6	6	71	171	119	28	10	18	181	121	17	17	17	186	120	15	15	15	185	106	13	13	13	175	103	16	11	11
7	5	119	230	189	88	18	22	237	190	74	21	21	238	186	58	19	19	232	167	41	16	16	212	156	63	16	16
8	4	163	261	244	153	22	25	263	243	138	73	24	257	235	119	58	21	249	214	98	71	18	214	194	109	20	19
9	3	201	269	280	212	104	48	262	275	197	138	50	249	263	176	119	40	228	243	151	132	68	188	204	146	71	20
10	2	230	261	295	259	167	104	235	286	244	198	104	219	263	222	176	106	194	243	194	143	132	142	197	170	132	78
11	1	249	226	290	284	222	167	202	275	275	244	167	176	253	252	218	167	151	224	224	184	178	83	172	178	131	131
12		255	183	265	298	265	183	150	245	284	245	245	114	218	261	218	218	71	185	234	234	200	17	131	209	178	178

June 21

AM	PM	0°	46° E	46° SE	46° S	46° SW	46° W	56° E	56° SE	56° S	56° SW	56° W	66° E	66° SE	66° S	66° SW	66° W	76° E	76° SE	76° S	76° SW	76° W	90° E	90° SE	90° S	90° SW	90° W
4	8	0	16	6	2	2	2	17	5	2	2	2	18	5	2	2	2	18	5	2	2	2	18	4	1	1	1
5	7	40	108	60	2	14	14	120	58	11	11	11	120	58	11	11	11	121	55	10	10	8	116	48	8	8	8
6	6	86	186	129	34	21	21	196	126	57	17	17	186	119	57	15	13	185	119	52	15	12	152	103	13	12	12
7	5	132	238	193	92	25	25	245	186	116	21	21	242	176	116	21	18	234	174	55	20	16	210	156	55	16	16
8	4	175	266	245	154	48	28	264	238	170	76	23	257	236	170	116	23	226	214	98	68	18	184	180	98	17	17
9	3	212	272	278	211	109	33	261	261	214	139	104	249	242	214	163	24	215	234	143	127	20	139	192	133	70	18
10	2	240	260	293	255	169	109	234	257	242	193	163	221	246	210	178	25	189	234	184	178	22	82	185	156	78	18
11	1	258	231	293	284	222	169	204	251	251	212	212	210	241	219	210	57	138	215	210	210	99	161	172	164	121	121
12		264	189	263	294	263	189	156	234	276	234	251	118	212	251	212	118	78	178	219	219	178	18	121	164	121	121

184

BTU/FT² Hourly Total Insolation, $I_{D\theta} + I_d$, On Surfaces Tilted With Respect To The Horizontal

Date	Solar Time AM	PM	0°	46° E/W	46° SE/SW	46° S	46° SW/SE	46° W/E	56° E/W	56° SE/SW	56° S	56° SW/SE	56° W/E	66° E/W	66° SE/SW	66° S	66° SW/SE	66° W/E	76° E/W	76° SE/SW	76° S	76° SW/SE	76° W/E	90° E/W	90° SE/SW	90° S	90° SW/SE	90° W/E
July 21	4	8	0	0	0	0	0	0	0	0	0	0	0	0	0	0	0	0	0	0	0	0	0	0	0	0	0	0
	5	7	27	80	44	11	11	11	86	45	10	10	10	90	44	9	9	9	91	43	8	8	8	89	39	6	6	6
	6	6	72	168	117	30	19	18	177	119	18	18	18	181	117	16	16	16	180	112	14	14	14	170	100	12	12	12
	7	5	119	225	185	88	24	22	232	186	74	22	21	232	181	58	20	20	226	172	41	18	17	206	150	61	15	14
	8	4	163	256	239	151	44	27	257	237	136	25	25	251	229	117	40	23	238	214	96	20	20	208	184	106	21	16
	9	3	201	264	274	208	105	29	258	269	193	67	27	244	257	173	106	25	223	237	147	77	21	184	197	142	72	17
	10	2	230	253	290	254	166	77	237	280	239	159	52	215	263	227	168	53	187	238	189	139	22	140	192	165	120	18
	11	1	248	224	285	283	219	131	200	271	268	195	108	170	248	258	220	114	136	218	216	181	75	82	167	191	165	18
		12	254	182	260	293	260	182	150	241	278	241	150	114	214	269	214	114	75	181	225	181	75	18	127	200	127	18
Aug 21	5	7	25	104	75	16	12	12	102	78	11	11	11	95	80	10	10	10	95	80	9	9	9	95	75	7	7	7
	6	6	72	187	159	73	19	19	183	164	65	18	18	178	165	56	16	16	147	161	45	14	14	147	148	28	11	11
	7	5	114	230	223	140	29	24	235	227	119	21	21	231	225	119	19	19	211	216	104	17	17	193	193	78	14	14
	8	4	146	243	264	202	92	29	241	265	193	67	27	232	259	179	106	24	216	245	160	82	21	224	214	126	51	15
	9	3	166	240	282	251	155	52	237	280	242	182	62	225	258	227	168	106	245	231	206	147	36	165	187	164	113	16
	10	2	218	204	279	282	211	131	200	271	271	211	118	182	255	258	221	162	218	218	245	189	82	120	167	196	106	17
		12	225	161	254	293	254	161	133	232	285	232	133	101	220	269	220	101	67	181	245	181	67	18	127	214	127	18
Sept 21	7	5	25	94	84	36	8	8	102	91	36	8	8	108	95	34	7	7	109	97	32	6	6	109	95	28	5	5
	8	4	72	207	179	111	15	14	183	190	111	14	14	188	195	108	13	13	177	194	102	11	11	177	184	89	9	9
	9	3	114	211	236	236	68	18	205	245	194	119	17	203	248	178	34	13	196	243	168	16	15	176	224	147	11	11
	10	2	146	196	262	271	192	35	192	268	242	182	62	137	266	267	102	167	138	256	211	82	13	80	210	193	113	12
	11	1	166	168	261	283	211	159	155	262	273	232	78	109	255	279	258	220	80	241	254	147	15	12	173	223	168	12
		12	173	124	236	283	236	124	102	232	285	232	102	78	220	279	220	78	51	202	265	202	51	12	168	233	168	12
Oct 21	8	4	20	79	87	53	6	6	86	92	57	6	6	91	102	59	5	5	94	105	57	5	5	92	104	57	4	4
	9	3	60	140	178	138	42	12	148	192	145	35	11	151	200	148	26	10	150	202	147	18	9	141	194	138	7	7
	10	2	92	148	225	201	104	35	149	231	210	98	53	146	236	213	153	105	139	235	210	79	10	122	221	195	60	8
	11	1	112	126	225	240	161	151	119	233	250	159	71	109	234	258	201	118	95	229	248	142	11	72	210	230	120	9
		12	119	86	204	253	204	86	71	207	263	207	71	54	204	266	204	54	36	195	261	195	36	9	173	241	173	9
Nov 21	9	3	12	45	63	49	12	4	49	70	54	11	4	52	75	57	10	3	53	78	58	9	3	53	78	58	6	2
	10	2	39	132	141	132	62	9	89	153	143	62	8	89	161	149	16	6	89	165	152	58	6	83	161	148	51	6
	11	1	58	78	164	194	114	11	76	175	209	118	10	66	182	217	118	8	66	183	203	115	8	57	195	196	106	7
		12	65	47	151	194	151	47	39	159	209	159	39	30	162	217	162	30	21	161	219	161	21	14	151	211	151	14
Dec 21	9	3	0	3	4	3	1	0	3	5	4	1	0	3	5	4	0	0	3	5	4	1	0	4	5	4	0	0
	10	2	19	55	91	86	39	8	55	101	95	40	7	57	108	101	95	7	57	111	104	40	6	55	110	103	37	5
	11	1	37	53	178	141	88	8	55	139	173	93	7	51	146	163	118	95	51	148	167	115	8	44	145	164	89	5
		12	43	32	122	159	122	32	26	130	173	130	26	21	135	182	135	21	14	136	186	136	14	5	130	182	130	5

aBased on data in Table 1, p 387, 1972 ASHRAE Handbook of Fundamentals; 0% ground reflectance, 1.0 clearness factor.

Table A.6. Btu/ft² Insolation on Inclined Surfaces at 64° North Latitude

Btu/ft² Hourly Total Insolation, $I_{D\theta} + I_d$, On Surfaces Tilted With Respect To The Horizontal

Date	Solar Time AM	PM	0°	54° E/W	54° SE/SW	54° S	54° SW/SE	54° W/E	64° E/W	64° SE/SW	64° S	64° SW/SE	64° W/E	74° E/W	74° SE/SW	74° S	74° SW/SE	74° W/E	84° E/W	84° SE/SW	84° S	84° SW/SE	84° W/E	90° E/W	90° SE/SW	90° S	90° SW/SE	90° W/E
Jan. 21	10	2	4	10	19	17	4	1	11	20	19	7	1	11	21	20	7	1	11	22	20	7	1	11	22	20	7	1
	11	1	12	24	64	72	42	4	24	69	77	44	3	24	72	80	45	3	23	72	81	45	3	22	73	81	44	2
	12		16	11	66	91	68	11	9	72	98	72	9	7	74	102	74	7	4	74	103	74	4	3	73	103	73	3
Feb. 21	8	4	4	27	30	17	2	2	29	33	19	1	1	30	34	19	1	1	31	35	19	1	1	31	35	19	1	1
	9	3	31	102	137	103	20	7	108	146	108	16	6	110	153	111	12	6	109	153	110	8	5	107	149	107	5	4
	10	2	55	114	187	170	75	9	116	196	178	72	9	114	200	181	67	8	108	198	178	60	7	104	194	173	55	6
	11	1	71	90	195	212	130	11	86	202	220	130	10	79	203	223	126	8	70	198	219	119	7	63	193	213	113	7
	12		77	48	173	225	173	48	37	177	235	177	37	26	175	237	175	26	14	169	232	169	14	7	162	226	162	7
Mar. 21	7	5	18	86	77	30	5	5	92	82	29	5	5	96	85	29	4	4	96	85	27	4	4	95	84	25	3	3
	8	4	54	163	173	101	10	10	171	181	102	9	9	173	185	99	8	8	170	182	94	7	7	166	178	89	7	7
	9	3	87	184	229	171	45	13	187	236	172	32	12	184	237	169	19	10	176	231	160	9	9	169	224	153	7	8
	10	2	112	171	252	227	109	14	166	257	229	98	13	157	254	224	85	12	144	244	213	68	10	134	234	203	58	9
	11	1	129	134	247	262	170	25	122	248	265	162	13	107	241	259	150	12	89	228	246	133	10	77	217	235	121	9
	12		134	83	218	274	218	83	64	215	277	215	64	44	205	271	205	44	22	189	258	189	22	9	177	246	177	9
Apr. 21	5	7	5	24	13	2	2	2	26	14	2	2	2	27	14	2	2	2	27	14	1	1	1	27	13	1	1	1
	6	6	37	130	92	15	10	10	137	95	63	14	9	140	95	54	12	12	139	93	63	7	7	137	90	37	6	6
	7	5	76	204	169	70	15	15	204	173	128	16	14	204	171	116	22	18	194	165	158	10	10	185	159	145	11	9
	8	4	117	229	227	136	18	18	230	230	189	45	16	227	225	224	24	18	211	214	203	56	14	205	214	188	47	11
	9	3	144	207	264	197	67	19	225	264	239	112	18	229	242	255	88	18	211	240	233	116	17	209	227	216	109	13
	10	2	169	168	278	246	132	20	196	273	270	174	27	207	242	266	152	27	183	221	243	126	14	183	204	225	163	14
	11	1	184	117	268	278	191	59	149	260	239	203	90	168	255	281	174	90	101	207	233	181	38	81	186	225	163	31
	12		190	117	238	289	238	117	90	225	281	225	90	62	206	281	206	62	31	181	243	181	31	13	177	225	177	13
May 21	4	8	11	12	15	13	5	5	14	14	10	4	4	15	14	10	4	4	13	13	9	3	3	12	12	8	3	3
	5	7	42	192	135	29	13	13	131	95	66	14	16	133	129	54	14	11	131	62	39	12	9	185	58	38	11	8
	6	6	79	234	196	86	18	18	197	134	129	16	19	221	186	116	18	16	191	121	94	15	11	211	115	80	13	11
	7	5	117	252	244	148	21	21	236	194	190	21	21	232	229	170	24	19	221	174	145	16	13	209	164	128	14	13
	8	4	152	239	273	204	23	23	239	236	235	53	21	252	252	213	48	17	222	230	186	56	14	183	198	167	15	15
	9	3	182	226	283	249	82	28	209	264	264	176	22	242	254	242	88	18	207	209	213	116	18	138	186	193	95	16
	10	2	205	188	273	278	143	83	163	274	274	225	47	251	199	251	148	20	168	207	222	168	38	81	147	201	147	16
	11	1	219	138	273	289	199	116	107	225	225	225	107	199	206	266	199	73	101	222	181	38	147	201	147	40
	12		224	138	244	268	244	138	107	274	281	274	107	73	199	251	199	73	38	168	222	168	38	40	147	201	147	40
June 21	3	9	4	14	20	10	2	2	15	28	9	2	2	15	25	8	2	2	15	22	7	2	2	15	20	6	1	1
	4	8	27	78	83	16	10	10	82	81	15	9	9	83	69	13	8	8	80	69	11	7	7	80	64	10	6	6
	5	7	60	149	144	34	16	16	154	141	74	15	17	155	134	112	13	13	147	123	36	11	13	147	115	33	10	13
	6	6	96	206	201	91	21	21	210	196	133	21	19	207	225	187	17	19	191	171	88	16	15	191	160	73	15	16
	7	5	132	242	245	150	24	24	242	238	187	23	21	236	248	164	20	21	209	205	137	18	16	206	191	128	16	17
	8	4	166	258	273	204	28	28	252	258	238	56	23	240	246	222	25	19	222	222	167	53	17	179	204	157	17	17
	9	3	195	247	282	247	88	39	211	268	268	168	24	247	249	242	99	21	197	211	199	111	19	136	199	181	92	18
	10	2	217	230	272	275	167	92	167	273	273	229	55	229	242	229	145	23	162	198	202	159	92	80	177	177	80	18
	11	1	231	193	272	284	201	177	113	258	267	242	113	78	194	251	194	78	40	161	211	161	40	18	139	189	139	18
	12		235	146	244	284	244	146	113	222	267	222	113	78	194	242	194	78	40	161	211	161	40	18	139	189	139	18

186

BTU/FT² Hourly Total Insolation, $I_{D\theta} + I_d$, On Surfaces Tilted With Respect To The Horizontal

Orientation note: For each tilt angle the five sub‑columns are read, in the AM, as E, SE, S, SW, W; in the PM the same columns are read as W, SW, S, SE, E. (Column labels below show the AM orientation with the PM orientation in parentheses.)

Date	AM	PM	0°	54° E(W)	SE(SW)	S	SW(SE)	W(E)	64° E(W)	SE(SW)	S	SW(SE)	W(E)	74° E(W)	SE(SW)	S	SW(SE)	W(E)	84° E(W)	SE(SW)	S	SW(SE)	W(E)	90° E(W)	SE(SW)	S	SW(SE)	W(E)
July 21	4	8	13	44	16	6	6	6	47	16	5	5	5	48	15	5	5	5	48	13	4	4	4	47	13	4	4	4
	5	7	44	124	69	14	14	14	129	68	13	13	13	130	65	11	11	11	128	61	10	10	10	125	57	9	9	9
	6	6	81	188	132	30	19	19	193	131	17	17	17	192	126	16	16	16	186	118	13	13	13	179	111	12	12	12
	7	5	118	229	192	86	23	23	231	189	72	21	21	226	181	56	18	18	215	169	38	16	16	205	159	28	14	14
	8	4	152	248	238	146	25	25	244	234	131	23	23	234	223	113	20	20	216	205	91	17	17	203	192	77	16	16
	9	3	182	244	268	201	83	26	234	260	186	55	24	218	245	166	25	21	195	223	141	18	18	178	207	124	17	17
	10	2	204	223	268	245	142	31	205	267	230	116	25	182	248	208	87	22	154	222	181	56	19	135	203	162	36	17
	11	1	218	186	240	273	197	85	162	253	258	174	50	133	230	236	146	22	100	201	207	114	19	80	181	187	93	17
	12	12	223	138	240	282	240	138	108	221	267	221	108	74	195	245	195	74	39	164	216	164	39	18	143	195	143	18
Aug. 21	5	7	6	26	15	3	3	3	28	15	3	3	3	29	15	2	2	2	29	15	2	2	2	29	15	2	2	2
	6	6	39	133	88	16	12	12	129	90	13	11	11	131	90	11	10	10	130	87	8	8	8	128	84	7	7	7
	7	5	77	187	159	69	21	21	193	163	61	16	16	193	161	52	17	17	187	155	42	14	12	181	149	35	13	11
	8	4	113	218	211	132	24	24	219	219	123	23	19	214	214	112	20	19	188	203	97	16	14	194	193	87	14	13
	9	3	144	221	253	190	69	26	216	252	182	47	21	205	244	169	24	21	150	228	150	44	16	176	216	138	16	14
	10	2	168	202	267	237	130	31	190	262	229	110	22	172	250	215	87	22	96	231	194	62	17	135	216	179	46	15
	11	1	183	165	259	268	186	66	146	250	260	169	50	123	234	244	148	50	64	211	222	122	33	79	195	205	105	15
	12	12	188	117	231	278	231	117	91	218	270	218	91	63	198	255	198	63	34	174	232	174	34	15	156	215	156	15
Sept. 21	7	5	16	71	63	25	6	6	75	67	25	5	5	78	69	24	5	5	78	69	23	4	4	78	69	21	4	4
	8	4	51	147	155	92	12	12	153	162	92	11	11	155	165	90	10	10	152	163	85	8	8	148	159	81	8	7
	9	3	83	170	211	159	21	21	173	218	159	32	16	170	218	156	20	17	162	212	147	10	12	155	205	141	10	9
	10	2	108	170	253	213	66	24	156	239	213	94	19	147	236	209	94	20	135	231	198	66	14	125	218	189	36	11
	11	1	124	126	232	246	161	27	116	233	248	154	22	102	234	243	142	23	85	213	230	126	17	73	203	220	94	12
	12	12	129	81	206	258	206	81	63	202	260	202	63	44	193	254	193	44	24	178	241	178	24	11	166	230	166	11
Oct. 21	8	4	2	13	15	9	1	1	15	16	9	1	1	15	17	10	1	1	17	17	10	1	1	15	17	10	1	1
	9	3	26	85	114	86	18	10	90	122	91	15	9	91	126	93	8	8	89	125	92	7	7	89	125	90	7	7
	10	2	50	102	167	152	68	12	103	175	159	65	12	93	177	161	61	11	93	173	159	55	10	93	173	155	8	9
	11	1	65	83	178	193	119	20	79	184	201	119	16	64	181	203	116	14	58	176	200	109	11	58	176	195	55	11
	12	12	71	45	159	207	159	45	35	162	215	162	35	24	155	217	155	24	13	149	213	149	13	8	149	208	149	8
Nov. 21	10	2	3	11	20	18	7	1	12	20	20	8	1	12	21	21	8	1	12	20	21	8	1	12	23	21	7	1
	11	1	12	24	63	70	41	4	22	68	76	43	4	21	70	79	44	3	23	71	80	44	3	22	70	79	43	3
	12	12	17	11	66	89	66	11	7	71	96	71	7	3	73	100	73	3	4	73	101	73	4	1	72	100	72	1
Dec. 21	11	1	0	1	3	3	2	0	1	3	4	2	0	1	3	4	2	0	1	3	4	2	0	1	3	4	2	0
	12	12	2	1	10	14	10	1	1	12	16	12	1	1	12	16	12	1	1	12	17	12	1	0	12	17	12	0

[a]Based on data in Table 1, p. 387, *1972 ASHRAE Handbook of Fundamentals*; 0% ground reflectance, 1.0 clearness factor.

187

Table A.7. Solar Position and Insolation Values for Various North Latitudes

Date	AM	PM	Alt@Lat 24	32	40	48	56	64	Az@Lat 24	32	40	48	56	64	Dir.Norm 24	32	40	48	56	64	Horiz 24	32	40	48	56	64
Jan 21	7	5	4.8	1.4	—	—	—	—	65.6	65.2	—	—	—	—	71	1	—	—	—	—	10	0	—	—	—	—
	8	4	16.9	12.5	8.1	3.5	—	—	58.3	56.5	55.3	54.6	—	—	239	203	142	37	—	—	83	56	28	4	—	—
	9	3	27.9	22.5	16.8	11.0	5.0	—	48.8	44.0	44.0	42.6	41.8	—	288	269	239	185	78	—	151	118	83	46	11	—
	10	2	37.2	30.6	23.8	16.9	9.9	2.8	36.1	30.9	30.9	29.4	28.5	28.1	308	295	274	239	170	22	204	167	127	83	39	2
	11	1	43.6	36.1	28.4	20.7	12.9	5.2	19.6	17.5	16.0	15.1	14.5	14.1	317	306	289	261	207	81	237	198	154	107	58	12
	12		46.0	38.0	30.0	22.0	14.0	6.0	0.0	0.0	0.0	0.0	0.0	0.0	320	310	294	267	217	100	249	209	164	115	65	16
Feb 21	7	5	9.3	7.1	4.8	2.4	—	—	74.6	73.5	72.7	72.2	—	—	158	121	69	12	—	—	35	22	10	1	—	—
	8	4	22.3	19.0	15.4	11.6	7.6	3.4	67.2	64.4	62.2	60.7	59.4	58.7	263	247	224	188	129	35	116	95	73	49	25	4
	9	3	34.4	29.9	25.0	19.7	14.2	8.6	57.6	53.4	50.2	47.7	45.9	44.8	298	288	274	251	214	147	187	161	132	100	65	31
	10	2	45.1	39.1	32.8	26.2	19.4	12.6	44.2	39.4	35.9	33.1	31.5	30.3	314	306	295	278	250	199	241	212	178	139	98	55
	11	1	53.0	45.6	38.1	30.5	22.8	15.1	25.0	21.4	18.9	17.2	16.1	15.3	321	315	305	290	266	222	276	244	206	165	118	71
	12		56.0	48.0	40.0	32.0	24.0	16.0	0.0	0.0	0.0	0.0	0.0	0.0	324	317	308	293	270	228	288	255	216	173	126	77
Mar 21	7	5	13.7	12.7	11.4	10.0	8.3	6.5	83.8	81.9	80.2	78.7	77.5	76.5	194	185	171	153	128	95	60	54	46	37	28	18
	8	4	27.2	25.1	22.5	19.5	16.2	12.7	76.8	73.0	69.6	66.8	64.4	62.6	267	260	250	236	215	185	141	129	114	96	75	54
	9	3	40.2	36.8	32.8	28.2	23.3	18.1	67.9	62.1	57.3	53.4	50.3	48.1	295	290	282	270	253	227	212	194	173	147	118	87
	10	2	52.3	47.3	41.6	35.4	29.0	22.3	54.8	47.5	41.9	37.8	34.9	32.7	309	304	297	287	272	249	266	245	218	187	151	112
	11	1	61.9	55.0	47.7	40.3	32.7	25.1	33.9	26.8	22.6	19.8	17.9	16.6	315	311	305	295	282	260	300	277	247	212	172	129
	12		66.6	58.0	50.0	42.0	34.0	26.0	0.0	0.0	0.0	0.0	0.0	0.0	317	313	307	298	284	263	312	287	257	220	179	134
Apr 21	6	6	4.7	6.1	7.4	8.6	9.6	10.0	100.6	99.9	98.9	97.8	96.8	96.0	89	66	41	1	0	27	7	14	20	27	35	37
	7	5	18.3	18.8	18.9	18.6	18.0	17.9	94.9	92.2	89.9	87.8	86.5	85.7	203	206	206	162	122	133	83	86	87	85	93	96
	8	4	32.0	31.5	30.3	28.5	26.1	24.5	89.0	84.0	79.3	74.9	70.9	67.5	256	255	252	219	201	208	160	156	137	131	129	122
	9	3	45.6	43.9	41.3	37.8	33.9	29.3	81.9	74.2	67.2	61.2	56.3	52.3	280	278	274	248	239	228	227	220	207	191	175	152
	10	2	59.0	55.7	51.2	45.8	39.9	33.5	71.8	60.3	51.4	44.6	39.7	36.0	292	290	286	264	259	248	278	267	250	228	207	187
	11	1	71.1	65.4	58.7	51.5	44.1	36.5	51.6	37.5	29.2	24.0	20.7	18.4	298	295	292	274	268	260	310	297	277	252	230	219
	12		77.6	69.6	61.6	53.6	45.6	37.6	0.0	0.0	0.0	0.0	0.0	0.0	299	297	293	280	275	263	321	307	287	260	240	224
May 21	5	7	—	—	1.9	5.2	8.5	11.6	—	—	108.4	108.1	106.8	106.1	—	—	1	41	93	51	—	—	1	9	21	21
	6	6	8.0	10.4	12.7	14.7	16.5	17.9	108.4	107.2	105.6	103.7	101.5	99.3	144	119	144	162	175	132	22	36	45	60	71	78
	7	5	21.2	22.8	24.0	24.6	24.8	24.5	103.2	100.1	96.6	92.9	89.3	85.9	216	211	216	216	219	185	98	107	114	118	119	132
	8	4	34.6	35.4	35.4	34.7	33.1	30.9	98.5	92.9	87.2	81.6	76.3	71.6	250	250	250	248	244	218	171	175	175	171	163	154
	9	3	48.3	48.1	46.8	44.3	40.9	36.8	93.6	84.7	76.0	68.3	61.6	56.1	267	269	267	269	259	239	233	233	227	219	201	182
	10	2	62.0	60.6	57.5	53.0	47.6	41.6	87.7	73.3	60.9	51.3	44.2	38.9	273	280	277	274	268	252	281	277	267	252	231	205
	11	1	75.5	72.0	66.2	59.5	52.3	44.8	76.9	51.9	37.1	29.7	24.9	20.9	277	285	280	279	273	261	311	307	293	274	249	219
	12		86.0	78.0	70.0	62.0	54.0	46.0	0.0	0.0	0.0	0.0	0.0	0.0	279	286	284	280	275	267	322	315	301	281	255	224
Jun 21	5	7	—	—	4.2	7.9	11.4	14.7	—	—	111.6	110.2	109.0	108.4	—	—	22	77	21	21	—	—	4	21	45	60
	6	6	9.3	12.2	14.8	17.2	19.3	21.0	111.3	110.2	108.4	106.8	104.7	103.2	97	131	155	172	185	93	19	40	60	77	86	96
	7	5	22.3	24.3	26.0	27.0	27.6	27.9	106.8	103.4	100.2	96.8	93.5	90.7	210	210	216	220	218	154	101	123	123	132	132	132
	8	4	35.5	36.9	37.4	37.1	36.0	34.0	103.4	96.8	90.7	84.7	79.7	74.9	242	245	246	246	243	193	171	182	182	180	175	166
	9	3	49.0	49.6	48.8	46.9	43.8	39.9	98.7	89.4	80.2	71.6	64.1	58.3	261	264	263	264	257	218	234	225	233	220	225	195
	10	2	62.6	62.2	59.8	55.8	50.7	44.9	92.2	78.0	65.8	55.4	47.9	42.2	274	272	272	269	265	239	280	272	272	246	269	217
	11	1	76.3	74.2	69.2	62.7	55.6	48.3	79.7	55.8	41.9	33.3	27.6	23.0	279	274	277	274	269	251	309	280	296	269	280	231
	12		89.4	81.5	73.5	65.5	57.5	49.5	0.0	0.0	0.0	0.0	0.0	0.0	281	279	279	275	271	258	319	315	304	287	264	235

Solar Position / BTU/FT² Hourly Total Insolation, $I_{DN} + I_d$

Solar Position / BTU/FT² Hourly Total Insolation, $I_{DN} + I_d$

Column groups — Solar Position: Altitude @ Lat., Azimuth @ Lat.; BTU/FT² Hourly Total Insolation: Direct Normal @ Lat., Horizontal Surface @ Lat. Each @ Lat. group has sub-columns for latitudes 24, 32, 40, 48, 56, 64.

Date	AM	PM	Alt 24	Alt 32	Alt 40	Alt 48	Alt 56	Alt 64	Az 24	Az 32	Az 40	Az 48	Az 56	Az 64	DN 24	DN 32	DN 40	DN 48	DN 56	DN 64	Hor 24	Hor 32	Hor 40	Hor 48	Hor 56	Hor 64
Jul 21	4	8	—	—	—	—	1.7	6.4	—	—	—	—	125.8	125.3	—	—	—	—	0	53	—	—	—	—	0	13
	5	7	—	—	2.3	5.7	9.0	12.1	—	—	115.2	114.7	113.7	112.4	—	—	2	43	91	128	—	—	0	10	27	44
	6	6	8.2	10.7	13.1	15.2	17.0	18.4	109.0	107.7	106.1	104.1	101.9	99.4	81	113	138	156	169	179	23	37	50	62	72	81
	7	5	21.9	23.1	24.3	25.1	25.3	25.0	103.8	100.6	97.2	93.5	89.7	86.0	195	203	208	211	212	211	98	107	114	118	119	118
	8	4	34.8	35.8	35.4	35.1	33.6	31.3	99.2	93.6	87.8	82.1	76.7	71.8	239	241	241	240	237	231	169	174	174	175	163	152
	9	3	48.4	48.4	47.2	44.8	41.4	37.3	94.5	85.1	76.7	68.8	62.0	56.3	261	261	259	256	252	245	231	231	225	215	201	182
	10	2	62.1	60.9	57.9	53.5	48.2	42.2	89.0	74.3	61.7	51.9	44.6	39.2	272	271	269	266	261	253	278	278	265	250	230	204
	11	1	75.7	72.4	66.7	60.1	52.9	45.4	79.2	51.3	37.9	29.0	23.7	20.2	278	277	275	271	265	257	307	302	290	272	248	218
	12		86.6	78.6	70.6	62.6	54.6	46.6	0.0	0.0	0.0	0.0	0.0	0.0	280	279	276	272	267	259	317	311	298	279	254	223
Aug 21	5	7	—	—	—	—	2.0	4.6	—	—	—	—	109.2	108.8	—	—	—	—	1	29	—	—	—	—	1	6
	6	6	5.0	6.5	7.9	9.1	10.2	11.0	101.3	100.5	99.5	98.3	97.0	95.5	35	59	81	99	112	123	7	14	21	28	34	39
	7	5	19.3	19.1	19.3	19.0	18.5	17.7	95.6	92.7	90.0	87.2	84.5	82.2	186	190	191	190	187	181	82	85	87	85	82	77
	8	4	33.2	31.8	30.7	29.0	26.7	23.9	89.7	84.2	80.2	76.8	71.3	69.0	241	242	237	232	225	214	158	156	150	141	128	113
	9	3	45.9	44.3	41.8	38.3	34.3	29.6	83.1	75.0	69.6	65.4	56.7	55.6	265	263	260	254	246	234	223	216	205	189	168	144
	10	2	59.3	56.1	51.7	46.4	40.5	34.2	73.0	61.3	55.3	51.1	40.0	41.1	278	276	272	266	258	246	273	262	246	225	198	168
	11	1	71.6	66.0	59.3	52.2	44.8	37.2	53.2	38.4	32.9	29.5	20.9	21.7	284	282	278	272	264	252	304	292	273	248	218	183
	12		78.3	70.3	62.3	54.3	46.3	38.3	0.0	0.0	0.0	0.0	0.0	0.0	286	284	280	274	266	254	315	302	282	256	225	188
Sept 21	7	5	13.7	12.7	11.4	10.0	8.3	6.5	83.8	81.9	80.2	78.7	77.5	76.5	173	163	149	131	107	77	57	51	43	35	25	16
	8	4	27.2	25.1	22.5	19.5	16.2	12.7	76.8	73.0	69.6	66.8	64.4	62.6	248	240	230	215	194	163	136	124	109	92	72	51
	9	3	40.2	36.8	32.8	28.2	23.1	18.1	67.9	62.1	57.3	53.4	50.3	48.1	278	272	263	251	233	206	205	188	167	142	114	83
	10	2	52.3	47.3	41.6	35.4	29.0	22.3	54.8	47.5	41.9	37.8	34.9	32.7	292	287	280	269	253	229	258	237	211	181	146	108
	11	1	61.9	55.0	47.7	40.3	32.7	25.1	33.4	26.8	22.6	19.8	17.9	16.6	299	294	287	278	263	240	291	268	239	205	166	124
	12		66.0	58.0	50.0	42.0	34.0	26.0	0.0	0.0	0.0	0.0	0.0	0.0	301	296	290	280	266	244	302	278	249	213	173	129
Oct 21	7	5	7.1	6.8	4.5	2.0	—	—	74.1	73.1	72.3	71.9	—	—	138	99	48	4	—	—	32	19	7	0	—	—
	8	4	20.0	18.7	15.0	11.2	7.1	3.0	66.7	64.0	61.9	60.2	59.1	58.5	247	229	204	165	104	17	111	90	68	44	20	2
	9	3	34.1	29.5	24.5	19.3	13.8	8.1	57.1	53.0	49.8	47.4	45.7	44.6	284	273	257	233	193	122	180	155	126	94	60	26
	10	2	44.7	38.7	32.4	25.7	19.0	12.1	43.8	39.1	35.6	33.1	31.3	30.2	301	293	280	262	231	176	234	204	170	133	92	50
	11	1	52.5	45.1	37.6	30.0	22.3	14.6	24.7	21.1	18.7	17.1	16.0	15.2	309	302	291	274	248	201	268	236	199	157	112	65
	12		55.5	47.5	39.6	31.5	23.5	15.5	0.0	0.0	0.0	0.0	0.0	0.0	311	304	294	278	253	208	279	247	208	166	119	71
Nov 21	7	5	4.9	1.5	—	—	—	—	65.8	65.4	—	—	—	—	67	2	—	—	—	—	10	0	—	—	—	—
	8	4	17.0	12.7	8.2	3.6	—	—	56.6	54.4	54.7	53.8	—	—	232	196	136	36	—	—	82	55	28	5	—	—
	9	3	28.0	22.6	17.0	11.2	5.2	—	46.1	44.1	42.7	42.0	41.9	—	282	263	232	179	76	—	150	118	82	46	12	—
	10	2	37.3	30.8	24.0	17.1	10.1	3.0	33.2	31.0	29.5	28.8	28.5	28.1	303	289	268	233	165	23	203	166	126	83	39	3
	11	1	43.8	36.2	28.6	20.9	13.1	5.4	17.6	16.1	15.1	14.7	14.5	14.2	312	301	283	255	201	79	236	197	153	107	58	12
	12		46.2	38.2	30.2	22.2	14.2	6.2	0.0	0.0	0.0	0.0	0.0	0.0	315	304	288	261	211	97	247	207	163	115	65	17
Dec 21	7	5	3.2	—	—	—	—	—	62.6	—	—	—	—	—	30	—	—	—	—	—	3	—	—	—	—	—
	8	4	14.9	10.3	5.5	1.9	—	—	53.8	53.0	53.0	52.1	—	—	225	176	89	14	—	—	71	41	14	5	—	—
	9	3	25.5	19.8	14.0	8.0	1.9	—	43.6	41.9	41.9	40.9	40.5	—	281	257	217	140	5	—	137	102	65	27	0	—
	10	2	34.3	27.6	20.7	13.6	6.6	—	31.2	29.4	29.4	28.5	28.2	—	304	288	261	214	113	—	189	150	107	63	19	—
	11	1	40.4	32.7	25.0	17.3	9.5	1.8	16.4	15.2	15.2	14.7	14.4	13.7	314	301	280	242	164	4	221	180	134	86	37	0
	12		42.6	34.6	26.6	18.6	10.6	2.6	0.0	0.0	0.0	0.0	0.0	0.0	317	304	285	250	180	16	232	190	143	94	43	2

aBased on data in Table 1, p. 387, *1972 ASHRAE Handbook of Fundamentals*; 0% ground reflectance, 1.0 clearness factor.

Table A.8. Values for Day-Long Insolation of Flat, Inclined and Vertical Surfaces with Various Wall Azimuths

BTU/FT² Daily Total Insolation, $I_{DB} + I_d$

DATE & DECL	DEC N. LAT	TOTAL I_{DN}	HORIZ	Σ = (L - 10°) E&W	SE&SW	S	Σ = L E&W	SE&SW	S	Σ = (L + 10°) E&W	SE&SW	S	Σ = (L + 20°) E&W	SE&SW	S	VERT E&W	SE&SW	S
Jan. 21 δ = -20°	24	2767	1622	1586	1864	1984	1525	1978	2174	1440	2034	2301	1350	2030	2360	752	1367	1767
	32	2459	1288	1221	1652	1839	1161	1743	2008	1097	1784	2118	1019	1776	2166	630	1323	1779
	40	2181	948	889	1417	1660	849	1497	1809	812	1539	1905	754	1535	1945	536	1264	1726
	48	1711	597	560	1105	1360	542	1169	1477	513	1200	1551	485	1196	1579	378	1065	1477
	56	1127	283	280	721	934	277	765	1011	267	788	1058	250	787	1075	215	748	1044
	64	306	45	50	199	269	48	212	290	46	220	303	42	220	306	39	218	304
Feb. 21 δ = -10.6°	24	3036	1998	1953	2178	2276	1871	2239	2396	1778	2241	2446	1656	2178	2425	911	1294	1476
	32	2873	1724	1638	2024	2188	1566	2076	2300	1478	2067	2345	1362	2010	2322	844	1359	1643
	40	2640	1413	1307	1830	2061	1250	1868	2163	1172	1863	2202	1089	1808	2177	749	1365	1731
	48	2331	1080	983	1588	1882	941	1628	1972	885	1621	2004	818	1566	1978	627	1302	1719
	56	1987	741	701	1326	1639	672	1359	1716	643	1352	1742	599	1306	1717	209	1182	1598
	64	1432	399	410	949	1231	401	974	1286	382	970	1303	352	938	1283	329	906	1252
Mar. 21 δ = 0.0°	24	3078	2270	2207	2362	2427	2118	2352	2455	1996	2290	2411	1842	2162	2298	973	1119	1023
	32	3012	2084	1970	2253	2378	1870	2245	2403	1743	2172	2359	1604	2055	2245	948	1229	1276
	40	2917	1852	1696	2125	2308	1600	2106	2330	1491	2040	2285	1370	1925	2173	912	1323	1484
	48	2781	1578	1415	1963	2209	1336	1947	2228	1253	1887	2183	1140	1773	2075	862	1388	1632
	56	2586	1269	1143	1771	2067	1095	1762	2084	1020	1703	2040	942	1594	1937	790	1405	1701
	64	2296	932	888	1536	1856	854	1526	1870	807	1472	1830	737	1380	1737	686	1311	1657
Apr. 21 δ = +11.9°	24	3036	2455	2393	2441	2457	2288	2352	2375	2143	2222	2227	1973	2036	2017	1088	882	488
	32	3077	2390	2256	2394	2443	2127	2308	2356	1980	2165	2206	1810	1987	1993	1034	1032	764
	40	3092	2274	2063	2324	2412	1939	2235	2320	1793	2094	2169	1623	1913	1956	1050	1176	1022
	48	3076	2107	1848	2232	2360	1732	2135	2265	1593	2005	2114	1442	1818	1903	1054	1298	1263
	56	3025	1893	1628	2110	2283	1525	2026	2187	1406	1892	2038	1270	1710	1831	1044	1405	1457
	64	2982	1643	1442	1985	2177	1359	1905	2081	1259	1771	1936	1134	1603	1737	1045	1484	1594
May 21 δ = +20.3°	24	3032	2556	2494	2465	2447	2381	2325	2286	2228	2149	2071	2052	1927	1799	1030	733	246
	32	3112	2582	2430	2459	2454	2291	2316	2284	2122	2124	2064	1926	1904	1787	1074	879	469
	40	3159	2552	2298	2416	2441	2141	2273	2264	1960	2079	2041	1759	1850	1761	1103	1031	723
	48	3254	2483	2143	2374	2417	1986	2218	2235	1812	2030	2009	1627	1789	1729	1159	1196	983
	56	3341	2373	1981	2310	2374	1839	2159	2187	1682	1964	1962	1498	1729	1682	1214	1349	1218
	64	3471	2237	1859	2236	2312	1739	2090	2124	1584	1888	1899	1413	1657	1624	1292	1499	1437
June 21 δ = +23.45°	24	2994	2573	2509	2451	2421	2393	2293	2331	2241	2099	1992	2059	1865	1699	1023	664	203
	32	3083	2635	2477	2460	2435	2331	2295	2234	2152	2085	1990	1946	1849	1691	1070	811	371
	40	3179	2647	2380	2438	2435	2212	2272	2225	2018	2056	1975	1811	1810	1672	1122	964	610
	48	3313	2626	2252	2413	2421	2082	2235	2204	1897	2022	1951	1695	1764	1645	1196	1141	875
	56	3438	2562	2112	2365	2388	1955	2185	2166	1776	1965	1912	1573	1708	1606	1267	1303	1121
	64	3651	2488	2026	2310	2343	1881	2132	2118	1703	1900	1863	1511	1643	1559	1376	1470	1356

BTU/FT² Daily Total Insolation, $I_{DH} + I_d$

DATE & DECL	DEC N. LAT	TOTAL I_{DN}	HORIZ	Σ = (L-10°)			Σ = L			Σ = (L+10°)			Σ = (L+20°)			VERT		
				E & W	SE & SW	S	E & W	SE & SW	S	E & W	SE & SW	S	E & W	SE & SW	S	E & W	SE & SW	S
July 21 δ = +20.5°	24	2932	2526	2464	2432	2413	2352	2291	2251	2201	2115	2036	2025	1894	1766	1013	716	246
	32	3012	2558	2407	2428	2421	2268	2284	2249	2099	2092	2029	1903	1872	1754	1057	859	459
	40	3062	2535	2281	2387	2409	2123	2242	2230	1941	2047	2006	1744	1818	1727	1087	1007	703
	48	3158	2474	2132	2347	2386	1973	2189	2201	1799	1999	1974	1613	1759	1694	1145	1169	957
	56	3241	2373	1971	2284	2342	1828	2129	2153	1668	1932	1926	1482	1697	1647	1198	1318	1185
	64	3373	2248	1854	2210	2281	1729	2061	2089	1572	1856	1863	1398	1624	1588	1277	1567	1402
Aug. 21 δ = +12.1°	24	2863	2408	2347	2388	2402	2243	2296	2316	2099	2165	2168	1931	1980	1959	979	852	471
	32	2902	2352	2218	2343	2388	2089	2253	2296	1942	2109	2145	1772	1930	1934	1005	993	735
	40	2916	2244	2033	2273	2355	1907	2180	2258	1759	2037	2105	1589	1855	1893	1020	1130	979
	48	2899	2087	1823	2180	2300	1704	2079	2200	1562	1945	2047	1411	1758	1836	1023	1245	1207
	56	2849	1883	1605	2057	2219	1499	1967	2118	1377	1830	1966	1239	1649	1760	1013	1347	1393
	64	2808	1646	1422	1930	2109	1335	1844	2008	1231	1707	1861	1105	1539	1662	1015	1421	1521
Sept. 21 δ = 0.0°	24	2878	2195	2134	2281	2341	2047	2269	2366	1928	2207	2323	1778	2082	2213	933	1074	992
	32	2808	2014	1903	2170	2288	1805	2159	2309	1680	2086	2264	1544	1972	2154	907	1175	1226
	40	2709	1788	1635	2038	2211	1540	2017	2229	1432	1950	2183	1313	1830	2074	868	1258	1416
	48	2568	1521	1358	1871	2102	1279	1853	2118	1196	1792	2072	1085	1680	1967	815	1307	1545
	56	2368	1220	1088	1674	1949	1038	1661	1962	964	1601	1918	887	1496	1819	740	1298	1595
	64	2074	892	833	1430	1726	797	1417	1736	750	1364	1696	682	1275	1608	633	1210	1532
Oct. 21 δ = -10.7°	24	2869	1928	1884	2102	2197	1804	2162	2314	1712	2164	2364	1593	2103	2345	870	1251	1442
	32	2696	1655	1571	1943	2100	1499	1991	2208	1412	1982	2252	1300	1927	2231	798	1303	1589
	40	2455	1348	1241	1741	1962	1185	1777	2060	1108	1771	2098	1027	1718	2075	700	1295	1655
	48	2154	1022	925	1498	1775	884	1535	1860	830	1527	1891	765	1475	1867	583	1226	1626
	56	1805	689	648	1227	1516	618	1257	1587	591	1251	1611	549	1208	1589	465	1093	1480
	64	1239	357	358	839	1087	348	860	1136	330	856	1152	302	828	1134	282	800	1107
Nov. 21 δ = -19.9°	24	2706	1609	1573	1845	1962	1512	1955	2146	1427	2007	2268	1337	2001	2323	741	1340	1729
	32	2405	1280	1212	1633	1816	1153	1721	1979	1088	1759	2085	1009	1749	2129	621	1297	1741
	40	2128	943	883	1398	1635	842	1475	1779	804	1513	1870	746	1508	1907	528	1237	1687
	48	1668	596	557	1087	1336	538	1149	1449	508	1178	1518	480	1172	1544	373	1041	1442
	56	1094	284	278	706	913	275	749	987	264	769	1032	247	768	1046	211	728	1015
	64	302	47	51	198	266	49	210	287	46	217	299	42	218	302	302	215	300
Dec. 21 δ = -23.45°	24	2625	1475	1438	1730	1853	1383	1855	2059	1303	1925	2205	1221	1938	2286	676	1353	1808
	32	2348	1135	1081	1515	1705	1035	1620	1888	978	1678	2017	915	1689	2086	577	1315	1795
	40	1978	783	738	1247	1480	708	1336	1634	678	1388	1741	629	1399	1796	454	1192	1646
	48	1444	446	421	910	1137	406	975	1250	390	1012	1326	368	1020	1363	288	934	1304
	56	748	157	155	472	620	153	508	679	146	530	717	136	535	735	116	517	723
	64	24	2	2	15	21	2	16	23	2	17	24	2	18	24	2	17	24

[a]Based on data in Table 1, p. 387, 1972 ASHRAE *Handbook of Fundamentals*; 0% ground reflectance, 1.0 clearness factor.

Table A.9. Solar Angle of Incidence on Inclined Surfaces at 24° North Latitude

Solar Angle Of Incidence On Surfaces Tilted With Respect To The Horizontal

Sub-column labels for each tilt: **E / SE / S / SW / W** (AM); the same columns read **W / SW / S / SE / E** (PM).

Jan 21

Solar Time AM/PM	0°	14° E/W	14° SE/SW	14° S	14° SW/SE	14° W/E	24° E/W	24° SE/SW	24° S	24° SW/SE	24° W/E	34° E/W	34° SE/SW	34° S	34° SW/SE	34° W/E	44° E/W	44° SE/SW	44° S	44° SW/SE	44° W/E	90° E/W	90° SE/SW	90° S	90° SW/SE	90° W/E
7 / 5	85.2	72.5	72.2	62.4	81.7	90.0	63.6	62.9	75.9	90.0	90.0	54.8	53.8	72.6	90.0	90.0	46.3	45.0	69.8	90.0	90.0	24.8	21.1	65.7	90.0	90.0
8 / 4	73.1	61.4	59.5	49.3	67.8	85.1	53.4	49.9	62.0	79.9	85.1	45.9	40.4	58.5	83.2	90.0	39.3	31.4	56.0	86.8	90.0	35.5	21.3	58.8	90.0	90.0
9 / 3	62.1	52.1	48.1	37.1	55.9	72.0	45.8	38.2	48.4	66.2	72.0	40.6	28.2	44.5	69.2	89.1	37.0	18.3	42.2	72.8	89.1	48.4	28.1	54.4	90.0	90.0
10 / 2	52.8	45.6	39.0	27.3	46.4	61.8	42.0	29.3	35.5	53.0	61.8	40.2	19.7	30.6	55.1	76.2	40.4	11.0	28.2	58.7	83.8	62.0	38.1	50.0	82.9	90.0
11 / 1	46.4	43.7	34.1	23.2	40.1	53.5	43.2	26.3	24.8	40.8	52.4	44.9	20.3	16.0	41.8	64.1	48.3	18.1	14.1	44.6	70.9	75.9	49.2	47.0	71.9	90.0
12	44.0	45.7	35.2	30.0	35.2	45.7	48.9	31.0	20.0	31.0	48.9	53.4	29.4	10.0	29.4	53.4	58.8	30.8	0.0	30.8	58.8	90.0	60.6	46.0	60.6	90.0

Feb 21

Solar Time AM/PM	0°	14° E/W	14° SE/SW	14° S	14° SW/SE	14° W/E	24° E/W	24° SE/SW	24° S	24° SW/SE	24° W/E	34° E/W	34° SE/SW	34° S	34° SW/SE	34° W/E	44° E/W	44° SE/SW	44° S	44° SW/SE	44° W/E	90° E/W	90° SE/SW	90° S	90° SW/SE	90° W/E
7 / 5	80.7	67.2	68.6	54.9	82.6	90.0	57.6	60.2	75.2	90.0	90.0	48.2	52.1	73.7	90.0	90.0	39.0	44.5	72.6	90.0	90.0	18.0	30.9	74.8	90.0	90.0
8 / 4	67.7	54.9	54.9	42.0	73.5	82.6	46.1	46.6	60.5	78.2	82.6	37.7	37.5	59.3	83.2	90.0	31.5	29.5	58.5	88.3	90.0	31.5	31.0	69.0	90.0	90.0
9 / 3	54.9	42.9	42.9	30.9	59.7	69.0	36.9	32.4	45.9	63.7	69.0	30.6	23.2	44.3	68.4	87.3	27.1	15.0	44.5	73.6	90.0	45.9	36.4	63.8	90.0	90.0
10 / 2	44.9	34.2	32.9	21.2	46.4	55.6	34.1	19.5	31.1	49.4	55.6	27.3	10.9	29.5	53.7	71.8	31.1	6.8	30.6	58.9	80.4	60.5	45.1	59.6	80.4	90.0
11 / 1	37.0	32.0	24.3	17.6	36.4	46.6	38.0	13.9	19.5	36.8	46.6	31.2	8.4	14.8	39.0	58.7	41.3	14.7	17.6	44.2	66.6	75.2	55.5	56.9	78.1	90.0
12	34.0	35.2	25.8	21.0	25.8	35.2	40.8	23.3	10.0	23.3	40.8	29.4	23.3	1.0	23.3	46.6	53.4	29.4	9.0	29.4	53.4	90.0	66.7	56.0	66.7	90.0

Mar 21

Solar Time AM/PM	0°	14° E/W	14° SE/SW	14° S	14° SW/SE	14° W/E	24° E/W	24° SE/SW	24° S	24° SW/SE	24° W/E	34° E/W	34° SE/SW	34° S	34° SW/SE	34° W/E	44° E/W	44° SE/SW	44° S	44° SW/SE	44° W/E	90° E/W	90° SE/SW	90° S	90° SW/SE	90° W/E
7 / 5	76.3	62.4	65.6	52.5	85.1	90.0	52.5	58.4	75.0	90.0	90.0	42.6	51.7	75.2	90.0	90.0	32.7	45.4	75.9	90.0	90.0	15.0	40.8	84.0	90.0	90.0
8 / 4	62.8	49.3	51.2	39.7	76.5	82.5	39.7	43.8	60.0	76.9	82.5	30.3	36.7	60.5	83.3	81.9	21.5	31.4	62.0	90.0	90.0	30.0	43.3	78.3	90.0	90.0
9 / 3	49.8	37.2	37.2	28.6	63.9	67.9	28.6	28.8	45.0	62.0	67.9	21.7	21.7	44.3	68.3	73.3	13.6	17.6	48.4	75.6	90.0	45.0	45.3	69.4	90.0	90.0
10 / 2	37.7	27.3	24.6	22.2	46.4	49.7	22.2	14.6	30.0	47.1	54.3	23.6	8.4	31.5	56.4	67.9	16.0	7.2	35.5	60.2	85.1	60.0	52.9	69.4	90.0	90.0
11 / 1	28.1	23.2	17.1	33.4	36.4	39.7	24.3	6.5	15.0	32.4	45.5	25.5	8.4	18.0	38.1	54.1	35.5	19.3	24.8	45.5	72.2	75.0	52.9	66.9	84.6	90.0
12	24.0	10.0	17.1	33.4	17.1	27.6	17.9	17.9	0.0	17.9	33.4	10.0	23.3	10.0	23.3	40.8	31.0	31.0	20.0	31.0	48.9	90.0	73.1	66.0	73.1	90.0

Apr 21

Solar Time AM/PM	0°	14° E/W	14° SE/SW	14° S	14° SW/SE	14° W/E	24° E/W	24° SE/SW	24° S	24° SW/SE	24° W/E	34° E/W	34° SE/SW	34° S	34° SW/SE	34° W/E	44° E/W	44° SE/SW	44° S	44° SW/SE	44° W/E	90° E/W	90° SE/SW	90° S	90° SW/SE	90° W/E
6 / 6	85.3	71.6	77.6	61.8	90.0	90.0	61.8	72.3	90.0	90.0	90.0	52.0	67.5	90.0	90.0	90.0	42.3	63.3	90.0	90.0	90.0	11.6	55.8	90.0	90.0	90.0
7 / 5	71.7	57.8	63.1	47.6	82.6	85.7	47.8	57.6	75.3	72.9	90.0	37.9	53.0	77.4	83.7	90.0	28.0	49.4	80.2	90.0	90.0	18.9	52.3	89.1	90.0	90.0
8 / 4	58.0	44.0	48.6	34.1	72.0	72.0	36.8	43.0	60.7	61.2	69.1	25.4	38.7	49.7	69.1	90.0	5.6	25.2	54.4	77.2	88.1	32.0	52.3	89.1	90.0	90.0
9 / 3	44.7	31.6	34.0	29.5	58.3	58.3	28.3	28.3	46.2	46.5	54.5	17.0	24.6	26.2	55.3	73.7	17.0	20.6	43.2	62.9	74.0	46.2	62.6	78.4	90.0	90.0
10 / 2	31.0	18.9	19.5	24.6	44.5	44.5	18.9	13.9	32.0	31.9	40.4	17.0	15.4	25.1	40.2	59.7	31.0	25.3	34.9	49.1	59.7	60.7	71.2	77.6	90.0	90.0
11 / 1	18.9	11.0	5.2	14.6	34.6	38.2	14.6	17.4	18.9	19.2	36.6	31.0	15.4	15.6	26.5	50.0	31.0	36.1	34.9	36.1	59.7	75.3	71.2	78.4	87.2	90.0
12	12.4	18.6	10.1	1.6	10.1	18.6	17.4	17.4	11.6	17.4	17.4	36.1	36.1	26.2	36.1	35.9	36.1	36.1	26.2	36.1	54.3	90.0	81.3	77.6	81.3	90.0

May 21

Solar Time AM/PM	0°	14° E/W	14° SE/SW	14° S	14° SW/SE	14° W/E	24° E/W	24° SE/SW	24° S	24° SW/SE	24° W/E	34° E/W	34° SE/SW	34° S	34° SW/SE	34° W/E	44° E/W	44° SE/SW	44° S	44° SW/SE	44° W/E	90° E/W	90° SE/SW	90° S	90° SW/SE	90° W/E
6 / 6	82.0	68.8	76.0	86.6	90.0	90.0	59.4	72.1	90.0	89.6	90.0	50.1	68.7	90.0	90.0	90.0	41.2	65.9	90.0	90.0	90.0	20.0	63.7	90.0	90.0	90.0
7 / 5	68.6	52.1	62.0	72.4	82.5	82.5	45.6	58.0	75.9	75.5	89.6	36.2	55.0	79.6	84.2	89.1	27.1	53.1	83.6	90.0	90.0	24.8	60.7	90.0	90.0	90.0
8 / 4	55.4	41.6	47.9	67.0	69.7	69.7	31.8	44.1	62.0	61.4	75.7	6.1	41.8	53.5	70.2	89.1	3.3	32.6	59.5	79.1	85.7	35.5	60.7	88.9	90.0	90.0
9 / 3	41.7	27.8	33.9	44.5	55.7	55.7	18.8	30.6	46.1	41.4	56.4	16.8	22.9	22.1	56.4	75.7	16.8	29.0	42.6	65.6	72.0	48.4	63.9	86.7	90.0	90.0
10 / 2	28.0	14.1	20.0	30.6	42.0	42.0	4.2	18.8	33.8	43.1	42.8	16.0	32.4	33.0	43.2	62.6	31.0	32.4	42.6	51.2	58.2	62.0	69.8	86.7	90.0	90.0
11 / 1	14.5	3.3	11.5	24.8	27.7	28.3	10.4	13.9	24.8	21.3	38.2	44.1	22.9	30.0	31.3	48.2	41.2	41.2	40.0	41.2	58.1	75.9	77.8	86.7	87.2	90.0
12	4.0	14.5	11.5	10.0	11.5	14.5	24.3	21.3	20.0	21.3	24.3	30.0	31.3	30.0	31.3	34.2	40.0	41.2	40.0	41.2	34.2	90.0	87.2	86.7	87.2	90.0

June 21

Solar Time AM/PM	0°	14° E/W	14° SE/SW	14° S	14° SW/SE	14° W/E	24° E/W	24° SE/SW	24° S	24° SW/SE	24° W/E	34° E/W	34° SE/SW	34° S	34° SW/SE	34° W/E	44° E/W	44° SE/SW	44° S	44° SW/SE	44° W/E	90° E/W	90° SE/SW	90° S	90° SW/SE	90° W/E
6 / 6	80.7	67.7	75.4	86.0	90.0	90.0	58.6	72.1	90.0	89.2	90.0	49.7	69.3	90.0	90.0	90.0	41.1	67.1	90.0	90.0	90.0	23.5	66.9	90.0	90.0	90.0
7 / 5	67.7	54.4	61.7	72.4	81.2	82.5	45.1	58.4	76.3	89.2	90.0	35.9	55.1	80.0	84.2	87.9	27.4	54.8	85.0	90.0	90.0	27.6	64.1	90.0	90.0	90.0
8 / 4	54.5	40.9	48.0	66.6	69.5	69.5	31.4	44.9	62.7	75.4	78.0	4.6	43.5	57.3	70.8	74.8	4.6	32.6	61.8	79.1	84.8	37.4	64.1	88.9	90.0	90.0
9 / 3	41.0	27.3	34.3	31.6	57.0	57.0	17.6	32.1	49.6	48.1	64.8	16.8	32.6	55.3	55.4	64.4	16.8	32.4	52.4	65.8	71.3	49.6	67.2	88.7	90.0	90.0
10 / 2	27.3	13.7	21.1	27.7	39.1	41.4	4.1	21.4	35.1	48.1	47.7	16.8	25.7	34.4	43.2	62.6	16.8	32.4	45.9	54.5	58.1	62.7	72.8	86.7	90.0	90.0
11 / 1	13.7	0.3	13.6	27.7	27.7	27.7	6.1	27.6	35.5	23.6	38.2	44.0	33.6	31.4	33.1	47.7	44.0	43.6	43.4	43.6	44.0	76.3	80.5	89.4	89.6	90.0
12	0.6	14.0	13.6	13.4	13.6	14.0	24.0	23.6	23.4	23.6	24.0	34.0	33.6	31.4	33.1	34.0	44.0	43.6	43.4	43.6	44.0	90.0	89.6	89.4	89.6	90.0

Solar Angle Of Incidence On Surfaces Tilted With Respect To The Horizontal

Date	AM	PM	0°	14° E/W	14° SE/SW	14° S/S	14° SW/SE	14° W/E	24° E/W	24° SE/SW	24° S/S	24° SW/SE	24° W/E	34° E/W	34° SE/SW	34° S/S	34° SW/SE	34° W/E	44° E/W	44° SE/SW	44° S/S	44° SW/SE	44° W/E	90° E/W	90° SE/SW	90° S/S	90° SW/SE	90° W/E
July 21	6	6	81.8	68.6	75.9	86.5	90.0	90.0	59.2	72.1	90.0	90.0	90.0	50.1	68.8	90.0	90.0	90.0	41.1	66.1	90.0	90.0	90.0	20.6	64.2	90.0	90.0	90.0
	7	5	68.6	55.1	62.0	72.6	80.7	82.2	45.5	58.1	76.0	89.5	79.0	36.1	55.1	79.8	90.0	90.0	27.1	53.4	83.8	90.0	90.0	25.3	61.2	90.0	90.0	90.0
	8	4	55.2	41.4	47.9	58.5	66.9	69.0	31.7	44.2	62.1	75.5	65.5	22.1	42.1	66.4	84.2	88.9	13.2	41.9	71.4	90.0	85.5	35.8	61.3	85.0	90.0	90.0
	9	3	41.6	27.7	33.9	44.5	52.9	55.6	17.8	30.9	48.6	61.5	51.9	8.1	30.6	53.8	70.3	75.5	3.9	31.1	59.9	79.2	71.9	48.6	64.5	81.4	79.2	71.9
	10	2	27.9	13.9	20.1	30.7	38.8	41.9	3.9	19.2	35.8	47.5	38.1	6.1	22.9	42.5	56.8	61.9	16.1	29.6	50.1	65.8	58.1	62.1	70.3	79.1	65.8	58.1
	11	1	14.3	2.7	8.2	17.9	24.9	28.2	10.3	14.5	25.3	34.0	24.2	20.1	23.4	34.0	43.5	48.1	30.1	33.0	43.2	53.1	44.6	76.0	78.2	78.2	53.1	44.6
		12	3.4	14.4	11.8	11.4	11.8	14.4	24.2	21.7	20.6	21.7	24.2	34.1	31.7	30.6	31.7	34.1	44.1	41.7	40.6	41.7	44.1	90.0	87.6	86.3	87.6	90.0
Aug 21	6	6	85.0	71.3	77.4	87.9	90.0	90.0	61.5	72.3	90.0	90.0	90.0	51.8	67.6	90.0	90.0	90.0	42.2	63.5	90.0	90.0	90.0	12.3	56.4	90.0	90.0	90.0
	7	5	71.5	57.6	63.0	73.4	82.5	85.4	47.6	57.6	75.4	88.1	81.8	37.7	53.1	77.8	83.7	90.0	27.9	49.7	80.5	88.2	90.0	19.3	53.0	89.8	90.0	90.0
	8	4	57.8	43.8	48.5	58.9	68.1	71.8	33.8	43.0	60.8	73.6	68.0	23.8	38.9	63.3	69.1	83.2	13.8	26.8	63.6	72.8	80.5	32.1	58.7	85.7	90.0	83.1
	9	3	44.1	30.3	34.0	44.2	53.6	58.0	20.5	28.4	46.3	59.1	54.3	11.1	24.3	50.2	55.0	69.3	5.0	21.3	54.8	58.7	66.8	46.3	63.2	81.4	77.8	69.9
	10	2	30.7	16.8	19.4	29.5	39.1	44.3	8.0	14.4	32.0	44.5	40.2	9.6	15.2	37.2	40.4	63.9	16.7	21.3	43.8	44.6	45.2	60.8	71.8	79.1	66.4	58.8
	11	1	18.4	5.5	5.5	14.9	24.5	30.1	13.3	7.7	19.3	30.0	26.5	21.9	16.0	26.8	26.9	48.1	30.9	29.5	35.5	29.4	32.0	75.4	71.8	78.3	66.4	60.9
		12	11.7	18.2	10.0	2.3	10.0	18.2	26.5	17.7	12.3	17.7	26.5	35.7	26.9	22.3	26.9	35.7	45.2	36.5	32.3	36.5	45.2	90.0	81.8	78.3	81.8	90.0
Sept 21	7	5	76.3	62.4	65.6	75.2	85.3	90.0	52.5	58.4	75.0	76.9	90.0	42.6	51.7	75.2	90.0	90.0	32.7	45.5	75.9	90.0	90.0	15.0	40.8	84.0	90.0	90.0
	8	4	62.8	49.3	51.2	60.5	70.7	76.9	39.7	43.8	60.0	62.0	86.1	30.3	36.7	60.5	68.3	81.9	21.5	31.4	75.0	89.8	90.0	30.0	45.3	73.3	90.0	90.0
	9	3	49.8	37.1	37.0	45.9	56.3	62.0	28.6	28.8	45.0	47.1	72.4	21.4	21.7	45.9	53.1	67.9	17.1	17.6	48.4	75.0	77.2	45.0	52.9	69.4	75.1	62.9
	10	2	37.1	23.2	24.0	31.5	42.1	49.7	24.3	14.6	30.0	32.4	58.7	20.7	6.8	31.5	38.3	54.1	23.6	9.5	24.8	60.2	62.9	60.0	52.9	66.9	60.2	48.9
	11	1	28.1	11.7	14.6	18.0	28.5	37.5	33.4	6.5	15.0	17.9	45.5	28.8	8.4	18.0	23.1	40.8	35.5	19.5	13.5	45.5	49.8	75.0	73.1	66.0	45.5	32.0
		12	24.0	24.0	17.1	10.0	17.1	27.6	33.4	17.7	13.2	17.7	33.4	35.7	23.3	20.0	23.3	35.7	48.9	31.0	20.0	31.0	48.9	90.0	73.1	66.0	73.1	90.0
Oct 21	7	5	80.9	67.5	68.8	77.4	87.9	90.0	57.9	60.3	75.3	90.0	90.0	48.5	52.2	73.6	83.2	85.5	39.3	44.5	72.5	88.2	90.0	18.2	30.4	74.3	90.0	90.0
	8	4	68.0	55.3	55.1	63.1	73.7	80.9	46.5	46.1	60.6	78.3	76.7	38.1	37.6	58.9	68.4	73.6	30.5	29.8	58.4	72.6	80.3	31.6	30.5	68.5	88.1	83.1
	9	3	55.9	44.6	42.3	49.2	59.8	67.9	37.3	32.7	46.0	63.8	63.7	31.4	23.4	44.2	53.7	62.0	27.6	15.0	44.3	58.2	66.8	46.7	34.6	59.1	77.8	70.8
	10	2	45.3	36.7	31.3	36.2	46.6	55.7	32.6	21.3	31.6	50.5	50.1	30.9	11.3	29.5	39.4	53.6	32.0	1.5	17.3	44.2	53.8	62.7	44.2	56.4	66.4	58.7
	11	1	37.5	36.9	24.8	20.5	34.8	44.1	41.2	23.7	16.8	35.8	41.6	46.9	24.9	10.5	24.9	46.9	41.6	29.4	9.5	29.4	53.6	75.5	66.4	55.5	44.2	29.4
		12	34.5	34.5	26.3	20.5	26.3	36.9	41.2	23.7	18.5	23.7	41.2	46.9	24.9	0.5	24.9	46.9	53.6	29.4	9.5	29.4	53.6	90.0	66.4	55.5	66.4	90.0
Nov 21	7	5	85.1	72.4	72.1	79.6	90.0	90.0	63.4	62.9	75.9	90.0	90.0	54.7	53.8	72.6	90.0	90.0	46.2	44.9	69.8	90.0	90.0	24.7	21.3	65.9	90.0	90.0
	8	4	73.0	61.3	59.4	66.1	76.8	85.1	53.3	49.8	61.9	79.8	80.2	45.8	40.3	58.5	83.2	82.5	39.1	31.1	56.1	86.8	86.8	35.4	28.3	59.9	90.0	86.5
	9	3	62.0	51.9	48.0	53.4	63.8	72.8	45.6	38.0	48.3	66.1	68.7	40.4	28.1	44.4	55.2	70.8	36.8	18.0	44.2	58.7	83.7	48.3	38.2	54.5	83.1	72.1
	10	2	52.7	43.4	38.9	42.3	51.8	62.1	43.0	29.1	35.4	40.7	57.8	40.0	20.1	31.6	41.7	64.0	40.3	18.0	28.2	44.6	70.8	61.9	49.3	50.1	72.1	60.7
	11	1	46.6	43.5	33.0	29.8	42.3	52.5	48.7	30.9	19.8	30.9	48.7	47.3	29.3	9.8	29.3	47.3	48.1	30.8	0.2	30.8	58.7	75.9	60.7	46.2	44.6	30.8
		12	43.8	43.8	35.0	29.8	35.0	45.5	48.7	30.9	19.8	30.9	48.7	53.2	29.3	9.8	29.3	53.2	58.7	30.8	0.2	30.8	58.7	90.0	60.7	46.2	60.7	90.0
Dec 21	7	5	86.8	74.5	73.5	80.5	90.0	90.0	65.7	64.1	76.3	90.0	90.0	57.2	54.7	72.4	90.0	90.0	49.1	45.5	68.9	90.0	90.0	27.6	17.8	62.6	90.0	90.0
	8	4	75.1	63.8	61.3	67.5	78.0	86.7	56.1	51.6	62.7	80.2	82.5	48.9	41.9	58.6	69.6	90.0	42.5	32.3	58.4	86.5	90.0	37.4	18.1	56.6	90.0	90.0
	9	3	54.9	54.9	54.5	55.3	65.5	74.9	48.9	48.9	57.9	67.2	70.8	44.0	30.5	44.9	56.1	77.8	40.5	20.5	41.8	72.6	85.2	49.6	25.5	51.1	83.1	72.1
	10	2	55.7	48.8	42.0	36.5	54.1	64.1	45.5	29.8	37.4	43.0	60.3	43.6	23.0	31.6	43.2	66.2	43.6	20.3	28.8	45.2	72.5	62.7	35.9	46.6	80.7	69.9
	11	1	49.6	46.7	37.5	33.4	44.0	55.2	51.8	29.8	27.6	33.9	51.8	48.0	31.6	13.5	31.6	55.9	50.9	32.0	14.3	32.0	60.9	76.3	47.2	43.6	69.9	58.6
		12	47.4	49.0	38.5	33.4	38.5	49.0	51.8	31.6	23.5	31.6	51.8	55.9	31.6	13.5	31.6	55.9	60.9	32.0	3.4	32.0	60.9	90.0	58.6	42.5	58.6	90.0

193

aBased on data in Table 1, p. 387, *1972 ASHRAE Handbook of Fundamentals*; 0% ground reflectance, 1.0 clearness factor.

Table A.10. Solar Angle of Incidence on Inclined Surfaces at 32° North Latitude

Solar Angle Of Incidence On Surfaces Tilted With Respect To The Horizontal

Date	Solar Time AM	PM	0°	22° E/W	22° SE/SW	22° S	32° E/W	32° SE/SW	32° S	42° E/W	42° SE/SW	42° S	52° E/W	52° SE/SW	52° S	90° E/W	90° SE/SW	90° S
Jan. 21	7	5	88.6	68.7	68.0	79.6	59.9	58.8	75.9	51.2	49.7	72.6	43.1	41.0	69.8	24.8	20.3	65.2
	8	4	77.5	56.5	50.5	66.2	52.0	46.3	60.5	45.1	36.7	58.5	39.2	27.4	56.0	35.5	16.9	57.4
	9	3	67.5	45.5	38.4	53.9	47.1	35.5	45.0	43.2	25.5	44.5	40.6	15.2	42.2	48.4	22.5	50.0
	10	2	59.4	41.2	29.0	41.5	37.2	19.2	31.5	37.0	16.3	30.6	39.0	19.1	46.8	62.0	32.6	43.8
	11	1	53.9	33.5	30.0	34.4	31.5	33.9	28.2	44.5	29.7	10.0	67.7	35.1	33.9	75.9	44.2	39.6
	12		52.0	38.8	30.0	30.0	58.5	35.2	20.0	62.8	33.9	10.0	67.7	35.1	33.9	90.0	56.1	38.0
Feb. 21	7	5	82.9	63.8	63.5	77.2	52.5	55.4	75.3	43.2	47.5	73.7	34.3	40.2	72.6	18.0	29.3	73.6
	8	4	71.2	50.5	50.5	62.9	43.1	41.5	60.5	33.2	35.2	59.0	29.2	25.4	58.5	31.5	26.9	65.9
	9	3	60.1	38.4	35.9	49.0	37.6	28.7	45.9	33.2	19.2	44.3	29.0	4.5	44.1	45.0	31.0	58.2
	10	2	50.9	29.0	25.2	35.9	37.2	19.0	31.5	37.0	9.7	29.5	39.0	19.1	30.6	60.2	50.1	53.2
	11	1	44.4	25.5	29.2	25.0	42.2	19.0	14.8	45.4	16.3	10.0	50.1	17.6	76.2	75.3	61.8	49.4
	12		41.9	30.0	30.0	20.0	50.9	28.2	10.0	56.5	29.7	0.0	62.8	33.9	33.9	90.0	61.8	48.0
Mar. 21	7	5	77.3	60.3	75.2	75.2	45.7	53.2	75.0	36.0	46.8	75.2	26.3	41.4	75.9	15.0	38.7	82.1
	8	4	64.9	46.2	60.5	62.1	35.1	38.4	60.0	26.5	31.8	60.5	19.4	27.0	62.0	30.0	36.9	74.6
	9	3	52.7	32.6	45.9	47.9	28.0	23.9	45.0	23.3	16.8	45.9	22.1	13.6	48.4	45.0	40.1	68.0
	10	2	41.2	20.8	31.5	34.1	27.7	10.8	30.0	28.3	1.8	31.5	32.3	9.4	64.6	60.0	47.3	62.7
	11	1	35.0	18.1	18.0	18.0	34.0	10.4	15.0	38.6	13.2	18.0	44.9	21.0	86.7	75.0	57.0	59.2
	12		32.0	22.1	10.0	10.0	44.0	23.4	90.0	50.9	28.2	10.0	58.5	35.2	58.5	90.0	68.0	58.0
Apr. 21	6	6	83.9	71.8	88.0	88.0	52.2	66.8	90.0	42.7	62.5	90.0	33.1	58.9	90.0	11.6	55.1	90.0
	7	5	71.2	53.7	73.8	73.5	49.3	53.5	73.5	29.3	48.0	77.6	19.3	45.2	80.2	18.9	49.9	84.9
	8	4	58.5	42.9	58.7	60.7	37.6	41.5	60.9	17.7	33.8	63.4	8.2	32.5	67.0	32.0	48.5	78.7
	9	3	46.1	28.5	44.2	45.9	25.4	22.8	45.0	19.7	17.8	46.8	13.3	27.7	49.8	46.2	51.0	73.8
	10	2	34.3	14.2	29.5	36.9	20.7	8.2	31.5	32.3	12.2	36.8	26.6	27.7	67.0	60.7	57.1	73.8
	11	1	24.6	4.0	14.8	23.4	25.1	22.3	37.5	19.7	17.8	59.6	40.6	39.5	83.1	75.3	57.1	69.6
	12		20.4	16.0	1.6	22.3	37.4	22.3	34.0	30.5	30.5	45.9	39.5	54.8	90.0	90.0	75.7	69.6
May 21	6	6	79.6	70.1	86.6	86.6	49.3	66.6	90.0	40.2	63.8	90.0	31.6	61.7	90.0	20.0	62.7	90.0
	7	5	67.2	56.1	72.6	71.9	35.9	52.5	75.9	26.5	50.1	79.6	17.5	49.1	83.6	24.8	58.1	90.0
	8	4	54.6	42.1	58.5	57.9	22.7	38.6	62.0	12.8	37.2	66.2	3.5	38.1	70.4	38.5	58.1	86.4
	9	3	41.9	28.0	44.5	43.9	10.4	25.3	53.3	3.6	26.2	53.5	10.9	30.3	59.3	48.4	59.1	81.9
	10	2	29.4	14.6	30.6	38.0	8.8	14.5	38.1	15.9	20.5	41.2	24.9	28.7	69.3	62.0	64.4	81.9
	11	1	18.0	4.6	17.6	36.5	20.7	25.5	38.1	29.0	20.5	60.5	39.0	34.2	83.9	75.9	72.1	79.0
	12		12.0	15.8	10.0	23.5	34.0	24.8	30.4	33.0	34.4	47.0	44.1	56.2	90.0	90.0	81.5	78.0
June 21	6	6	77.8	69.6	86.0	86.0	48.3	66.7	90.0	39.6	64.5	90.0	31.5	63.1	90.0	23.4	65.8	90.0
	7	5	65.7	55.9	72.2	72.6	35.6	53.0	76.3	25.9	51.3	80.5	17.8	51.0	85.0	27.6	61.5	90.0
	8	4	53.1	42.1	58.6	57.5	21.4	38.9	62.4	13.0	39.0	67.5	5.3	40.6	72.0	37.4	60.3	89.6
	9	3	40.4	28.5	44.4	43.9	8.5	26.6	54.5	1.6	23.9	61.3	12.5	31.2	70.4	49.6	62.4	85.2
	10	2	27.8	15.4	31.6	39.6	6.5	17.3	44.5	15.3	27.1	66.3	28.0	37.0	70.4	62.7	67.5	85.2
	11	1	15.8	11.1	19.6	36.5	19.6	27.6	36.5	29.1	36.4	56.2	38.8	37.0	79.4	76.3	74.9	81.4
	12		8.6	23.5	13.4	23.5	33.0	26.6	33.0	42.7	36.4	42.7	52.5	46.2	52.5	90.0	84.0	81.4

194

Solar Angle Of Incidence On Surfaces Tilted With Respect To The Horizontal

Date	Solar Time AM / PM	0°	22° (E/W)	22° (SE/SW)	22° (S)	22° (SW/SE)	22° (W)	32° (E/W)	32° (SE/SW)	32° (S)	32° (SW/SE)	32° (W)	42° (E/W)	42° (SE/SW)	42° (S)	42° (SW/SE)	42° (W)	52° (E/W)	52° (SE/SW)	52° (S)	52° (SW/SE)	52° (W)	90° (E/W)	90° (SE/SW)	90° (S)	90° (SW/SE)	90° (W)
Jul. 21	6 / 6	79.3	58.4	70.0	86.5	90.0	88.6	49.1	66.6	66.6	90.0	90.0	40.1	63.9	66.3	90.0	90.0	31.5	62.0	90.0	90.0	90.0	20.6	63.2	90.0	90.0	90.0
	7 / 5	66.9	45.4	56.1	72.6	87.6	85.9	35.8	52.6	52.6	87.6	88.5	26.3	50.3	52.1	88.5	90.0	17.5	49.4	83.8	90.0	90.0	25.3	58.7	90.0	90.0	90.0
	8 / 4	54.3	32.4	42.1	58.5	73.4	76.3	22.4	38.8	37.5	73.4	80.1	12.6	37.5	38.2	73.1	86.3	3.7	38.5	71.4	88.5	90.0	35.8	57.5	90.0	90.0	90.0
	9 / 3	41.6	19.7	28.1	44.5	59.1	63.6	10.0	25.6	25.6	59.1	66.7	3.0	24.1	24.7	59.1	74.3	10.9	30.9	59.1	83.1	90.0	48.6	59.7	85.5	90.0	90.0
	10 / 2	29.1	12.9	14.3	30.7	44.6	50.6	8.5	15.8	15.8	44.9	54.1	15.7	24.7	19.3	44.4	63.6	24.9	29.3	43.2	72.5	90.0	62.1	64.9	79.3	87.3	90.0
	11 / 1	17.6	5.2	14.9	17.9	30.2	37.5	20.5	25.1	12.3	30.8	41.2	29.5	34.7	24.7	30.2	56.9	38.9	44.5	28.4	62.8	90.0	76.0	72.6	79.6	87.3	90.0
	12	11.4	16.0	16.0	10.6	15.8	24.6	33.8	25.1	10.6	25.1	24.6	43.2	34.7	22.3	37.6	56.8	52.9	44.5	34.7	56.8	90.0	90.0	82.0	78.6	82.0	90.0
Aug. 21	6 / 6	83.5	61.9	71.6	87.9	90.0	90.0	52.2	66.8	73.4	90.0	90.0	42.5	62.6	77.8	90.0	90.0	32.9	59.1	90.0	90.0	90.0	12.3	55.7	90.0	90.0	90.0
	7 / 5	70.9	48.9	57.2	73.3	87.6	87.3	38.9	52.1	60.8	87.6	90.0	29.0	48.1	73.6	88.4	90.0	19.1	45.5	80.5	90.0	90.0	19.3	50.6	85.5	90.0	90.0
	8 / 4	58.2	36.3	42.8	58.8	73.4	80.1	26.4	37.5	48.3	73.4	80.8	16.7	34.0	63.6	74.1	90.0	7.5	32.9	67.3	88.4	90.0	32.2	49.2	85.5	90.0	90.0
	9 / 3	45.7	24.9	28.4	44.2	59.1	67.1	16.5	25.6	35.4	59.1	66.2	8.0	20.9	50.0	59.2	73.1	12.9	23.1	54.8	81.5	90.0	46.3	51.7	79.3	90.0	90.0
	10 / 2	33.9	17.6	14.0	30.2	44.6	54.1	9.0	15.7	22.4	44.9	51.6	12.8	18.3	37.2	44.8	63.6	19.3	21.1	43.8	72.5	90.0	60.8	57.7	74.5	87.3	90.0
	11 / 1	24.0	5.3	3.3	35.5	30.8	41.2	24.7	15.8	13.2	30.5	39.6	18.3	12.8	26.8	44.8	59.9	35.5	28.2	39.9	67.2	90.0	75.8	66.2	71.4	87.3	90.0
	12	19.7	29.2	15.8	32.3	22.4	29.2	37.0	22.4	12.3	22.4	24.4	45.6	30.8	22.3	37.6	56.9	54.6	39.9	30.8	45.6	90.0	90.0	76.2	70.3	76.2	90.0
Sept. 21	7 / 5	77.3	55.6	60.3	75.2	90.0	86.1	45.7	53.4	82.3	90.0	90.0	36.0	46.8	90.0	90.0	90.0	26.3	41.4	90.0	90.0	90.0	15.0	38.7	90.0	90.0	90.0
	8 / 4	64.9	44.1	46.2	60.5	76.5	73.1	35.1	38.4	67.4	88.5	90.0	26.5	31.8	82.3	90.0	90.0	19.4	27.0	88.2	90.0	90.0	30.0	36.9	90.0	90.0	90.0
	9 / 3	53.2	34.8	32.6	45.9	62.1	60.4	27.4	23.9	52.6	73.4	90.0	23.3	16.8	67.3	88.2	90.0	22.1	13.6	79.4	90.0	90.0	45.0	40.1	85.5	90.0	90.0
	10 / 2	42.7	29.7	20.8	31.5	47.9	52.6	27.9	13.0	37.2	59.9	90.0	28.1	11.2	52.6	73.9	90.0	24.9	9.4	66.6	86.7	90.0	60.0	57.0	79.7	90.0	90.0
	11 / 1	35.0	31.1	18.5	31.1	36.1	48.5	34.7	15.8	15.7	45.1	62.8	38.5	11.2	28.2	59.2	72.2	44.9	21.0	49.8	72.5	90.0	75.0	59.2	79.7	90.0	90.0
	12	32.0	32.0	10.0	10.0	22.1	38.2	44.0	23.4	12.3	23.4	44.0	50.9	28.2	22.3	43.5	56.8	58.5	35.2	33.3	58.5	90.0	90.0	68.0	58.0	68.0	90.0
Oct. 21	7 / 5	83.2	62.2	64.0	77.4	90.0	90.0	52.8	55.6	90.0	90.0	90.0	43.6	47.6	90.0	90.0	90.0	34.7	40.2	90.0	90.0	90.0	18.2	28.8	90.0	90.0	90.0
	8 / 4	71.3	52.0	50.8	63.1	79.5	90.0	41.6	41.0	83.8	90.0	90.0	36.1	31.4	88.2	90.0	90.0	29.7	10.8	88.0	90.0	90.0	31.6	26.4	90.0	90.0	90.0
	9 / 3	60.5	47.0	40.0	52.4	69.5	83.3	38.1	29.2	69.1	90.0	90.0	33.7	19.5	78.8	90.0	90.0	31.6	4.7	79.4	90.0	90.0	46.0	30.5	85.5	90.0	90.0
	10 / 2	50.7	41.2	29.0	36.2	52.4	66.7	37.7	19.4	55.0	90.0	90.0	35.7	10.2	58.7	82.2	90.0	30.4	19.1	63.3	90.0	90.0	60.6	39.1	76.4	90.0	90.0
	11 / 1	44.9	42.1	26.0	25.4	46.6	62.8	42.6	19.4	37.5	55.7	74.3	45.8	16.6	14.8	44.1	69.1	39.4	17.3	48.6	76.4	90.0	75.3	49.8	73.4	90.0	90.0
	12	42.5	42.5	30.4	20.5	30.4	55.0	51.3	20.6	10.5	28.6	46.9	56.8	29.8	0.5	33.9	63.0	63.0	33.9	9.5	33.9	90.0	90.0	61.5	47.5	68.0	90.0
Nov. 21	7 / 5	88.5	68.6	67.9	79.6	90.0	90.0	59.7	58.7	90.0	90.0	90.0	51.1	49.7	90.0	90.0	90.0	42.9	41.0	90.0	90.0	90.0	24.7	20.4	90.0	90.0	90.0
	8 / 4	77.3	59.4	55.8	66.1	82.5	90.0	51.8	46.2	85.3	90.0	90.0	44.9	36.6	88.2	90.0	90.0	39.0	27.4	88.0	90.0	90.0	35.4	17.1	90.0	90.0	90.0
	9 / 3	67.4	52.7	45.4	53.4	69.5	87.4	47.2	35.4	71.5	90.0	90.0	43.0	25.4	74.1	90.0	90.0	40.4	15.4	77.1	90.0	90.0	48.3	22.7	90.0	90.0	90.0
	10 / 2	59.2	49.4	37.9	42.0	57.3	72.7	46.9	28.4	58.2	74.0	90.0	46.0	19.4	60.5	74.1	90.0	46.7	12.1	63.6	80.1	90.0	61.9	32.7	77.8	90.0	90.0
	11 / 1	53.8	50.3	35.3	33.3	46.6	62.8	58.4	23.5	45.7	58.2	74.0	53.0	23.1	17.5	46.6	73.9	56.2	15.2	48.9	80.1	90.0	75.8	44.3	74.5	90.0	90.0
	12	51.8	55.0	38.6	29.8	38.6	55.0	58.4	35.0	19.8	35.0	58.4	62.6	33.8	9.8	33.8	62.6	67.6	35.1	0.2	35.1	90.0	90.0	56.2	47.5	68.2	90.0
Dec. 21	8 / 4	79.7	62.4	58.0	67.5	83.8	90.0	55.1	48.2	88.2	90.0	90.0	48.4	38.5	88.2	90.0	90.0	42.6	28.9	90.0	90.0	90.0	37.4	13.5	90.0	90.0	90.0
	9 / 3	70.2	56.1	48.2	55.3	71.1	85.9	50.8	38.2	72.5	90.0	90.0	46.7	28.2	74.5	90.0	90.0	43.9	18.2	76.9	90.0	90.0	49.6	19.9	88.7	90.0	90.0
	10 / 2	62.4	53.1	41.3	44.5	59.4	75.1	50.5	31.9	59.7	81.4	90.0	49.4	23.1	61.6	80.9	90.0	49.7	15.6	63.1	81.6	90.0	62.7	30.6	80.7	90.0	90.0
	11 / 1	57.3	53.1	38.9	36.5	49.3	65.7	54.2	31.8	47.8	60.9	74.0	55.9	26.4	19.6	47.8	75.9	58.7	21.8	49.3	81.6	90.0	76.3	44.3	76.4	90.0	90.0
	12	55.4	58.3	41.9	33.4	41.9	58.3	61.3	37.9	23.4	37.9	61.3	65.1	35.8	13.5	35.8	65.1	69.6	36.1	3.5	36.1	90.0	90.0	54.4	34.5	54.4	90.0

aBased on data in Table 1, p. 387, 1972 *ASHRAE Handbook of Fundamentals*; 0% ground reflectance, 1.0 clearness factor.

195

Table A.11. Solar Angle of Incidence on Inclined Surfaces at 40° North Latitude

Solar Angle Of Incidence On Surfaces Tilted With Respect To The Horizontal

Date	Solar Time AM	PM	0° S	30° E/W	30° SE/SW	30° S	30° SW/SE	30° W/E	40° E/W	40° SE/SW	40° S	40° SW/SE	40° W/E	50° E/W	50° SE/SW	50° S	50° SW/SE	50° W/E	60° E/W	60° SE/SW	60° S	60° SW/SE	60° W/E	90° E/W	90° SE/SW	90° S	90° SW/SE	90° W/E
Jan. 21	8	4	81.9	58.1	52.5	66.2	88.1	90.0	50.9	42.8	62.0	90.0	90.0	44.5	33.2	58.5	90.0	90.0	39.9	24.0	56.0	90.0	90.0	35.5	13.0	55.7	90.0	90.0
	9	3	73.2	54.4	43.2	53.5	75.0	90.0	49.6	33.2	48.4	76.5	90.0	46.0	23.2	44.5	78.5	90.0	43.9	13.2	42.7	80.8	90.0	48.4	16.8	46.4	89.0	90.0
	10	2	66.2	54.2	37.5	42.1	62.6	83.4	52.3	28.4	35.5	62.5	89.6	51.7	20.1	30.6	64.5	90.0	52.5	14.0	28.2	66.7	90.0	62.0	27.5	38.1	77.1	90.0
	11	1	61.6	57.8	37.2	33.4	51.3	73.1	58.6	30.8	24.8	50.4	78.0	60.5	26.4	17.6	50.8	83.1	63.4	25.3	14.1	52.7	88.4	73.9	39.2	32.3	64.8	90.0
	12		60.0	64.3	42.4	30.0	42.4	64.3	67.5	39.0	20.0	39.0	67.5	71.2	37.8	10.0	37.8	71.2	75.5	38.7	0.0	38.7	75.5	90.0	52.2	30.0	52.2	90.0
Feb. 21	7	5	85.2	56.8	59.1	77.2	90.0	90.0	47.5	50.8	75.2	90.0	90.0	38.5	43.1	73.7	90.0	90.0	30.0	36.3	72.6	90.0	90.0	18.0	28.0	72.7	90.0	90.0
	8	4	74.6	49.0	46.3	62.9	85.0	90.0	41.3	37.3	60.5	88.8	90.0	34.5	28.8	59.0	90.0	90.0	29.4	21.5	58.5	90.0	90.0	31.5	22.9	63.3	90.0	90.0
	9	3	65.2	44.5	35.2	49.0	71.1	89.0	39.6	25.9	45.9	74.3	89.0	36.4	15.7	44.3	78.0	90.0	35.8	8.3	44.5	81.9	90.0	45.9	25.4	54.5	81.9	90.0
	10	2	57.2	44.5	27.9	35.9	57.7	77.1	43.0	18.5	31.5	59.9	77.1	43.5	10.3	29.5	63.2	84.3	45.8	8.3	29.5	67.3	90.0	60.5	33.9	47.1	82.3	90.0
	11	1	51.9	48.6	27.4	25.0	45.0	66.0	50.0	22.0	18.9	46.0	66.0	53.6	20.3	14.8	50.9	72.6	58.0	23.0	14.8	52.5	78.4	75.2	45.0	41.9	69.7	90.0
	12		50.0	56.2	34.2	20.0	34.2	56.2	60.5	32.8	10.0	32.8	60.5	65.6	34.1	0.0	34.1	65.6	71.3	37.8	10.0	37.8	71.3	90.0	57.2	40.0	57.2	90.0
Mar. 21	7	5	78.6	49.1	55.1	75.2	90.0	90.0	39.4	48.2	75.0	90.0	90.0	29.8	42.2	75.2	90.0	90.0	20.7	37.6	75.9	90.0	90.0	15.0	36.8	80.4	90.0	90.0
	8	4	67.5	40.1	41.3	60.5	82.0	90.0	31.8	33.6	60.0	87.4	90.0	24.5	27.2	60.5	90.0	90.0	19.7	23.3	62.0	90.0	90.0	30.0	32.9	71.3	90.0	90.0
	9	3	57.2	34.6	28.4	45.9	67.7	83.4	29.6	19.5	45.0	72.5	90.0	27.1	12.2	44.3	77.8	90.0	28.0	10.8	48.4	83.3	90.0	45.0	34.8	63.0	85.3	90.0
	10	2	48.4	34.5	18.5	31.5	53.5	71.1	33.9	8.7	30.0	57.7	79.2	36.0	2.8	31.5	67.8	87.5	40.1	11.8	35.5	68.5	87.5	60.0	41.7	56.2	87.7	90.0
	11	1	42.3	39.6	17.9	18.0	39.8	59.2	42.8	14.8	15.0	43.0	66.4	47.0	17.8	18.0	57.7	73.9	53.3	24.1	24.8	53.7	79.1	75.0	51.6	51.6	63.0	90.0
	12		40.0	48.4	27.1	10.0	27.1	48.4	54.1	28.5	0.0	28.5	54.1	60.5	32.8	10.0	32.8	60.5	67.5	39.0	20.0	39.0	67.5	90.0	63.0	50.0	63.0	90.0
Apr. 21	6	6	82.6	53.0	66.2	88.0	90.0	90.0	43.2	61.7	90.0	90.0	90.0	33.5	58.0	90.0	90.0	90.0	24.1	55.2	90.0	90.0	90.0	11.6	54.3	90.0	90.0	90.0
	7	5	71.1	41.1	51.8	73.5	90.0	90.0	31.1	47.0	73.3	90.0	90.0	21.1	43.5	80.3	90.0	90.0	11.1	43.5	80.2	90.0	90.0	18.9	47.5	89.5	90.0	90.0
	8	4	59.7	30.6	37.5	58.9	78.8	90.0	21.3	32.3	60.7	85.1	90.0	13.1	29.4	63.4	90.0	90.0	11.6	29.5	67.0	90.0	90.0	32.4	44.5	80.5	90.0	90.0
	9	3	48.7	23.4	23.2	44.2	64.6	77.0	18.1	17.7	46.2	71.2	86.5	17.3	12.0	48.7	76.8	90.0	21.6	21.7	54.4	85.2	90.0	46.2	45.9	73.1	85.2	90.0
	10	2	39.3	21.5	9.2	29.5	50.2	64.5	24.3	4.2	32.0	56.5	73.6	28.8	12.0	36.8	63.4	82.8	35.5	21.7	43.2	70.8	90.0	60.7	51.5	67.0	89.2	90.0
	11	1	31.3	25.1	6.5	14.8	35.8	52.2	35.2	16.7	18.9	41.8	60.6	39.2	22.6	21.6	48.9	69.2	49.7	36.5	34.9	56.6	78.0	75.3	60.0	63.6	78.0	90.0
	12		28.4	40.4	21.6	1.6	21.6	40.4	47.6	27.1	11.6	27.1	47.6	55.5	34.6	21.6	34.6	55.5	63.9	43.0	31.6	43.0	63.9	90.0	70.3	61.6	70.3	90.0
May 21	5	7	88.1	61.1	78.3	90.0	90.0	90.0	52.5	75.6	90.0	90.0	90.0	43.5	74.1	90.0	90.0	90.0	36.6	71.6	90.0	90.0	90.0	24.8	69.8	90.0	90.0	90.0
	6	6	77.3	48.7	64.5	86.6	90.0	90.0	39.4	61.5	75.9	90.0	90.0	30.0	60.6	79.6	90.0	90.0	8.4	58.3	79.4	90.0	90.0	20.0	61.4	87.7	90.0	90.0
	7	5	66.0	36.3	50.0	72.6	90.0	90.0	26.1	45.5	62.0	90.0	90.0	17.6	45.9	63.4	90.0	90.0	5.9	46.2	66.2	90.0	90.0	24.8	55.5	80.5	90.0	90.0
	8	4	54.6	24.6	36.5	58.5	76.8	90.0	14.7	31.5	47.4	75.9	90.0	6.2	33.4	48.9	78.6	90.0	23.6	33.4	53.1	78.9	90.0	35.5	52.9	74.1	86.4	90.0
	9	3	43.2	15.2	22.4	44.5	63.0	84.5	9.8	20.7	33.7	62.0	84.5	11.5	22.2	33.3	64.9	86.8	29.3	29.2	42.1	64.4	89.2	48.4	54.1	68.9	86.5	90.0
	10	2	32.5	15.2	8.6	30.6	49.0	72.6	18.6	12.0	20.6	48.4	70.6	24.1	18.1	21.6	50.9	75.7	34.1	36.5	33.4	54.0	75.7	62.9	58.9	66.5	86.5	90.0
	11	1	23.8	23.9	7.2	17.6	34.9	63.0	31.0	16.7	8.9	35.5	59.9	39.2	26.6	9.2	38.6	66.1	48.1	36.5	22.2	42.6	72.0	76.9	66.5	62.0	74.6	90.0
	12		20.0	35.5	20.8	10.0	20.8	35.5	44.0	28.9	20.0	28.9	44.0	52.8	37.9	30.0	37.9	52.8	62.0	47.2	40.0	47.2	62.0	90.0	76.0	70.0	76.0	90.0
June 21	5	7	85.8	59.5	77.6	90.0	90.0	90.0	51.2	75.4	90.0	90.0	90.0	43.4	73.8	90.0	90.0	90.0	36.4	72.6	90.0	90.0	90.0	27.4	72.4	90.0	90.0	90.0
	6	6	75.2	47.1	64.0	86.0	90.0	90.0	38.2	61.7	76.3	90.0	90.0	29.9	60.2	85.0	90.0	90.0	22.5	59.8	79.4	90.0	90.0	23.6	64.3	90.0	90.0	90.0
	7	5	64.0	34.7	50.3	72.4	90.0	90.0	25.2	48.0	62.7	90.0	90.0	16.2	47.2	72.8	90.0	90.0	9.5	48.1	66.2	90.0	90.0	37.4	58.7	80.5	90.0	90.0
	8	4	52.6	22.6	36.6	58.6	76.0	90.0	12.6	34.7	49.6	76.6	90.0	2.7	35.4	61.7	90.0	90.0	7.4	35.4	52.4	78.9	90.0	49.6	57.5	83.6	90.0	90.0
	9	3	41.2	12.5	22.9	44.9	63.0	82.6	6.5	22.7	37.4	62.5	80.8	11.3	26.4	49.6	78.9	90.0	20.3	32.6	43.5	65.2	88.0	62.7	62.0	78.1	88.0	90.0
	10	2	30.2	6.5	9.6	31.6	48.8	70.9	20.3	16.6	21.6	49.6	78.2	24.9	23.7	37.4	63.6	78.2	34.0	39.2	34.9	54.9	76.9	76.3	69.2	74.6	86.8	90.0
	11	1	20.8	21.9	7.4	19.6	35.0	58.8	31.6	19.2	11.3	24.9	61.1	37.7	27.9	16.6	39.2	65.2	47.7	38.4	26.7	43.8	74.8	78.4	78.4	62.0	74.6	90.0
	12		16.6	33.9	21.4	13.4	21.4	33.9	42.8	30.3	23.4	30.3	42.8	52.0	39.6	33.4	39.6	52.0	61.4	49.2	43.4	49.2	61.4	90.0	78.4	73.4	78.4	90.0

196

Date	Solar Time		Solar Angle Of Incidence On Surfaces Tilted With Respect To The Horizontal																									
			0°	30°					40°					50°					60°					90°				
	AM	PM		AM E	SE	S	SW	W	E	SE	S	SW	W	E	SE	S	SW	W	E	SE	S	SW	W	E	SE	S	SW	W
				PM W	SW	S	SE	E	W	SW	S	SE	E	W	SW	S	SE	E	W	SW	S	SE	E	W	SW	S	SE	E
July 21	5	7	87.7	60.8	78.2	90.0	90.0	90.0	52.2	75.6	90.0	90.0	90.0	44.1	73.4	90.0	90.0	90.0	36.6	71.7	90.0	90.0	90.0	25.3	70.2	90.0	90.0	90.0
	6	6	76.9	48.4	64.4	86.5	90.0	90.0	39.2	61.6	90.0	90.0	90.0	30.4	59.6	90.0	90.0	90.0	22.5	58.6	90.0	90.0	90.0	20.6	61.9	90.0	90.0	90.0
	7	5	65.7	36.0	50.5	72.6	90.0	90.0	26.3	47.6	76.0	90.0	90.0	16.8	46.1	79.8	90.0	90.0	8.5	46.4	83.8	90.0	90.0	25.3	56.0	90.0	90.0	90.0
	8	4	54.2	24.3	36.5	58.5	76.7	84.2	14.3	33.8	62.1	84.6	90.0	4.6	33.7	66.4	90.0	90.0	6.1	36.1	71.4	90.0	90.0	35.8	53.5	88.2	90.0	90.0
	9	3	42.8	15.0	22.5	44.5	62.9	72.3	9.2	21.0	48.6	70.6	82.2	11.9	23.9	53.8	78.6	90.0	20.0	29.8	59.9	86.7	90.0	48.6	54.7	81.0	90.0	90.0
	10	2	32.1	14.6	8.9	30.7	48.9	60.0	18.2	12.6	35.8	56.6	69.6	25.4	20.9	42.5	64.7	79.3	34.0	30.2	50.1	73.1	89.0	62.1	59.4	75.4	90.0	90.0
	11	1	23.3	23.5	7.5	17.9	34.9	47.6	30.7	17.1	25.3	42.7	56.8	39.1	27.0	34.0	51.1	66.1	48.0	37.0	43.2	59.9	75.5	76.0	66.9	71.8	87.2	90.0
	12		19.4	35.2	20.9	10.6	20.9	35.2	43.7	29.1	20.6	29.1	43.7	52.7	38.2	30.6	38.2	52.7	61.9	47.5	40.6	47.5	61.9	90.0	76.4	70.6	76.4	90.0
Aug. 21	6	6	82.1	52.6	66.0	87.9	90.0	90.0	42.9	61.7	90.0	90.0	90.0	33.2	58.1	90.0	90.0	90.0	23.8	55.5	90.0	90.0	90.0	12.3	54.9	90.0	90.0	90.0
	7	5	70.7	40.7	51.7	73.4	90.0	90.0	30.7	47.0	75.4	90.0	90.0	20.7	43.7	77.8	90.0	90.0	10.7	42.0	80.5	90.0	90.0	19.3	48.2	90.0	90.0	90.0
	8	4	59.3	30.0	37.3	58.8	78.6	88.9	20.7	32.4	60.8	85.7	90.0	12.4	29.7	63.6	90.0	90.0	8.7	30.0	67.3	90.0	90.0	32.2	45.2	81.3	90.0	90.0
	9	3	48.2	22.7	23.0	44.2	64.4	76.6	17.4	17.8	46.3	71.1	86.2	16.8	17.3	50.0	78.1	90.0	21.3	21.8	54.8	85.3	90.0	46.3	46.6	73.7	90.0	90.0
	10	2	38.3	22.5	9.2	29.5	50.1	64.2	23.7	4.8	32.2	56.5	73.3	28.5	12.7	37.2	63.5	82.5	35.3	22.4	43.8	70.9	90.0	60.8	52.1	67.6	90.0	90.0
	11	1	30.7	29.4	7.7	14.9	35.7	51.8	34.8	12.8	19.3	41.8	60.3	41.7	21.6	26.8	49.0	69.0	49.5	31.1	35.5	56.9	77.8	75.4	60.5	63.7	82.3	90.0
	12		27.7	39.9	21.4	2.3	21.4	39.9	47.3	27.2	12.3	27.2	47.3	55.3	34.8	22.3	34.8	55.3	63.7	43.3	32.3	43.3	63.7	90.0	70.8	62.3	70.8	90.0
Sept. 21	7	5	78.6	49.1	55.1	75.2	90.0	90.0	39.4	48.2	75.0	90.0	90.0	29.8	42.2	75.2	90.0	90.0	20.7	37.6	75.9	90.0	90.0	15.0	36.8	80.4	90.0	90.0
	8	9	67.5	40.1	41.3	60.5	82.0	90.0	31.8	33.6	60.0	87.4	90.0	24.5	27.2	60.5	90.0	90.0	19.7	23.3	62.0	90.0	90.0	30.0	32.9	71.3	90.0	90.0
	9	3	57.2	34.6	28.4	45.9	67.7	83.4	29.6	19.5	45.0	72.5	90.0	27.1	12.2	45.9	77.8	90.0	28.0	10.8	48.4	83.3	90.0	45.0	34.8	63.0	90.0	90.0
	10	2	48.4	34.5	18.5	31.5	53.5	71.1	33.9	8.7	30.0	57.7	79.2	36.0	2.8	31.5	62.8	87.5	40.1	11.8	35.5	68.5	90.0	60.0	41.7	56.2	87.7	90.0
	11	1	42.3	3[illegible].6	17.9	18.0	39.8	59.2	42.8	14.8	15.0	43.0	66.4	47.6	17.8	18.0	47.8	73.9	53.6	24.7	24.8	53.7	81.6	75.0	51.5	51.6	75.2	90.0
	12		40.0	48.4	27.0	10.0	27.0	48.4	54.1	28.5	90.0	28.5	54.1	60.5	32.8	10.0	32.8	60.5	67.5	39.0	20.0	39.0	67.5	90.0	63.0	50.0	63.0	90.0
Oct. 21	7	5	85.5	57.2	59.3	77.4	90.0	90.0	47.9	51.0	75.3	90.0	90.0	39.0	43.2	73.6	90.0	90.0	30.5	36.3	72.5	90.0	90.0	18.2	27.6	72.4	90.0	90.0
	8	4	75.0	49.4	46.6	63.1	85.1	90.0	41.7	37.5	60.6	88.9	90.0	35.0	29.0	58.9	90.0	90.0	29.9	21.5	58.4	90.0	90.0	31.6	22.4	62.9	90.0	90.0
	9	3	65.5	45.0	35.6	49.2	71.3	89.3	40.1	25.7	46.0	74.4	90.0	36.9	16.0	44.2	78.0	90.0	35.9	6.9	44.3	81.9	90.0	46.0	25.0	54.1	90.0	90.0
	10	2	57.6	44.8	28.3	36.2	57.8	77.4	43.4	19.0	31.6	60.1	84.6	43.9	10.8	29.5	63.3	90.0	46.1	8.4	30.4	67.2	90.0	60.6	33.6	46.6	82.1	90.0
	11	1	52.4	49.0	27.9	25.4	45.3	66.3	50.9	22.5	18.2	46.1	72.3	54.0	20.5	14.8	48.6	78.6	58.3	23.0	17.3	52.5	85.1	75.3	44.7	41.4	69.5	90.0
	12		50.5	56.6	34.5	20.5	34.5	56.6	60.8	33.1	10.5	33.1	60.8	65.9	34.2	0.5	34.2	65.9	71.5	37.8	9.5	37.8	71.5	90.0	56.9	39.5	56.9	90.0
Nov. 21	8	4	81.8	57.9	52.4	66.1	88.0	90.0	50.7	42.7	61.9	90.0	90.0	44.3	33.1	58.5	90.0	90.0	39.0	23.9	56.1	90.0	90.0	35.4	13.2	55.8	90.0	90.0
	9	3	73.0	54.2	43.0	53.4	74.9	90.0	49.4	33.0	48.3	76.5	90.0	45.8	23.0	44.4	78.5	90.0	43.8	13.1	42.2	80.8	90.0	48.3	17.0	46.6	89.1	90.0
	10	2	66.0	54.0	37.3	42.0	62.4	83.3	52.1	28.2	35.4	63.0	89.5	51.6	19.9	30.5	64.5	90.0	52.4	13.8	28.2	66.7	90.0	61.9	27.6	38.4	77.2	90.0
	11	1	61.4	57.6	37.0	33.3	51.2	73.0	58.5	30.6	24.7	50.3	77.9	60.4	26.3	17.5	50.8	83.0	63.2	25.2	14.1	52.6	88.4	75.9	39.8	32.5	64.9	90.0
	12		59.8	64.2	42.2	29.8	42.2	64.2	67.3	38.9	19.8	38.9	67.3	71.1	37.7	9.8	37.7	71.1	75.4	38.7	0.2	38.7	75.4	90.0	52.3	30.2	52.3	90.0
Dec. 21	8	4	84.5	61.3	54.9	67.5	89.1	90.0	54.3	45.0	62.7	90.0	90.0	47.9	35.2	58.6	90.0	90.0	42.6	25.6	55.4	90.0	90.0	37.4	9.7	53.2	90.0	90.0
	9	3	76.0	57.8	46.1	55.3	76.4	90.0	53.0	36.1	49.6	77.4	90.0	49.3	26.2	44.9	78.8	90.0	47.0	16.3	41.8	80.5	90.0	49.6	14.3	43.8	87.0	90.0
	10	2	69.3	57.7	40.9	44.5	64.4	85.6	55.6	31.8	37.4	64.4	90.0	54.7	23.5	31.6	65.2	90.0	55.0	16.9	28.0	66.7	90.0	62.7	25.7	35.4	75.4	90.0
	11	1	65.0	61.0	40.6	36.5	53.7	75.7	61.5	34.0	27.6	52.2	80.1	63.0	29.0	19.6	51.9	84.8	65.3	26.8	14.3	53.0	89.7	76.3	38.2	29.0	73.2	90.0
	12		63.4	67.2	45.3	33.4	45.3	67.2	70.0	41.5	23.4	41.5	70.0	73.3	39.5	13.5	39.5	73.3	77.1	39.5	3.5	39.5	77.1	90.0	50.8	26.6	50.8	90.0

[a]Based on data in Table 1, p. 387, *1972 ASHRAE Handbook of Fundamentals;* 0% ground reflectance, 1.0 clearness factor.

Table A.12. Solar Angle of Incidence on Inclined Surfaces at 48° North Latitude

Solar Angle Of Incidence On Surfaces Tilted With Respect To The Horizontal

Date	AM	PM	0°	38° E/W	38° SE/SW	38° S/S	38° SW/SE	38° W/E	48° E/W	48° SE/SW	48° S/S	48° SW/SE	48° W/E	58° E/W	58° SE/SW	58° S/S	58° SW/SE	58° W/E	68° E/W	68° SE/SW	68° S/S	68° SW/SE	68° W/E	90° E/W	90° SE/SW	90° S/S	90° SW/SE	90° W/E
Jan 21	8	4	86.5	56.7	49.2	66.2	90.0	90.0	49.8	39.5	62.0	90.0	90.0	43.8	29.9	58.5	90.0	90.0	39.0	20.8	56.0	90.0	90.0	35.5	10.2	54.7	90.0	90.0
	9	3	79.0	56.0	41.1	53.1	90.0	90.0	51.6	31.1	48.4	90.0	90.0	48.4	21.1	44.5	82.2	90.0	46.6	11.3	42.2	83.7	90.0	48.4	11.2	43.7	87.6	90.0
	10	2	73.1	58.8	37.0	42.1	79.9	90.0	57.1	28.4	35.5	67.3	90.0	56.5	20.7	30.6	68.2	90.0	57.0	15.6	28.2	69.7	90.0	62.0	22.8	33.5	75.1	90.0
	11	1	69.3	64.7	39.0	33.4	55.7	82.6	65.4	33.0	24.8	54.3	86.8	66.8	29.0	17.6	54.3	90.0	69.0	27.9	14.1	55.6	90.0	75.9	35.8	25.4	62.2	90.0
	12		68.0	72.8	45.7	30.0	45.7	72.8	75.5	42.4	20.0	42.4	75.5	78.5	41.0	10.0	41.0	78.5	81.9	41.6	10.0	41.6	81.9	90.0	49.0	22.0	49.0	90.0
Feb 21	7	5	87.6	51.8	54.6	77.2	90.0	90.0	42.7	46.5	75.2	90.0	90.0	34.0	39.1	73.7	90.0	90.0	26.2	32.9	72.6	90.0	90.0	18.5	27.3	72.2	90.0	90.0
	8	4	78.4	46.9	42.3	62.9	90.0	90.0	39.8	35.0	60.5	90.0	90.0	34.7	22.9	59.0	90.0	90.0	31.9	13.4	58.5	90.0	90.0	31.6	19.9	61.1	90.0	90.0
	9	3	70.3	46.0	32.4	49.0	76.2	90.0	42.0	22.4	45.9	78.9	90.0	39.7	12.5	44.3	81.9	90.0	39.5	11.5	44.5	85.0	90.0	45.9	28.5	50.7	84.2	90.0
	10	2	63.8	49.4	27.3	35.9	62.6	87.4	48.6	18.5	31.5	64.5	90.0	49.3	11.7	29.5	67.1	90.0	51.5	26.3	30.6	70.5	90.0	60.5	28.5	41.4	75.5	90.0
	11	1	59.5	56.2	29.6	25.0	49.7	75.9	58.1	20.2	18.0	50.3	81.4	51.0	23.7	14.8	52.4	87.0	64.8	14.8	17.6	55.8	90.0	75.2	53.2	34.6	66.3	90.0
	12		58.0	65.3	38.1	20.0	38.1	65.3	69.2	36.8	10.0	36.8	69.2	73.7	37.9	90.0	37.9	73.7	78.5	41.0	10.0	41.0	78.5	90.0	49.0	32.0	53.2	90.0
Mar 21	7	5	80.0	43.0	50.2	75.2	90.0	90.0	33.5	43.6	75.0	90.0	90.0	24.4	38.2	75.2	90.0	90.0	16.2	34.5	75.9	90.0	90.0	15.0	35.0	78.9	90.0	90.0
	8	4	70.5	37.2	36.6	60.5	87.2	90.0	29.8	29.0	60.0	90.0	90.0	24.3	23.2	60.5	90.0	90.0	21.8	20.5	62.0	90.0	90.0	30.0	28.9	68.2	90.0	90.0
	9	3	61.8	40.1	24.6	45.9	72.9	90.0	32.6	15.4	45.0	77.2	90.0	31.7	8.2	45.9	81.8	90.0	33.6	9.8	48.4	86.7	90.0	45.0	29.4	58.3	84.2	90.0
	10	2	54.6	48.0	17.4	31.5	58.7	81.4	40.6	8.6	30.0	62.4	89.1	43.0	6.9	31.5	66.8	90.0	47.1	14.8	35.5	71.9	90.0	60.0	36.0	49.9	71.1	90.0
	11	1	49.7	58.2	20.9	18.0	44.8	69.5	51.3	19.0	15.0	47.6	76.1	55.8	21.9	18.0	51.8	82.9	61.2	28.1	24.8	57.1	89.9	75.0	46.3	44.1	58.3	90.0
	12		48.0	58.2	31.7	10.0	31.7	58.2	63.4	33.0	90.0	33.0	63.4	69.2	36.8	10.0	36.8	69.2	75.5	42.4	20.0	42.4	75.5	90.0	58.1	42.0	58.3	90.0
Apr 21	6	6	81.4	43.9	60.9	88.0	90.0	90.0	34.1	57.0	90.0	90.0	90.0	24.5	54.1	90.0	90.0	90.0	15.1	52.4	90.0	90.0	90.0	11.6	53.3	90.0	90.0	90.0
	7	5	71.4	33.5	46.6	73.5	90.0	90.0	23.6	42.4	75.3	90.0	90.0	13.7	39.8	77.6	90.0	90.0	4.6	39.2	80.2	90.0	90.0	18.9	45.0	86.9	90.0	90.0
	8	4	61.5	26.1	32.3	59.0	83.9	90.0	18.2	27.7	60.7	85.6	90.0	13.5	26.0	63.4	82.1	90.0	15.1	27.1	67.0	88.6	90.0	32.0	40.7	76.7	90.0	90.0
	9	3	52.2	24.6	18.2	44.2	69.7	90.0	22.4	13.1	46.2	70.4	90.0	24.2	14.5	49.7	67.4	90.0	29.3	21.1	54.4	74.2	90.0	46.2	40.7	67.7	89.8	90.0
	10	2	44.0	29.9	6.2	29.8	55.4	86.8	32.5	3.8	32.0	61.1	89.0	37.1	13.8	36.8	52.9	90.0	43.7	24.8	43.2	60.3	90.0	60.7	45.8	60.3	77.1	90.0
	11	1	38.5	39.2	20.0	16.2	41.6	74.7	45.7	17.1	18.9	46.4	76.4	48.1	26.2	26.2	38.5	78.5	58.1	33.7	34.9	53.3	86.7	75.3	54.5	55.3	65.2	90.0
	12		36.4	50.0	26.8	1.6	26.8	62.6	57.4	33.0	11.6	33.0	64.8	64.8	38.5	46.3	38.5	66.4	72.5	46.3	31.6	46.3	72.5	90.0	65.2	53.6	65.2	90.0
May 21	5	7	84.8	50.9	73.2	90.0	90.0	90.0	42.6	71.2	90.0	90.0	90.0	35.1	69.7	90.0	90.0	90.0	28.9	68.9	90.0	90.0	90.0	24.8	69.4	90.0	90.0	90.0
	6	6	75.3	38.8	59.4	86.6	90.0	90.0	29.7	57.1	90.0	90.0	90.0	21.3	55.9	90.0	90.0	90.0	14.9	55.9	90.0	90.0	90.0	20.8	59.8	90.0	90.0	90.0
	7	5	65.4	28.0	45.4	72.5	90.0	90.0	17.5	43.1	75.9	89.1	90.0	7.8	42.6	79.6	90.0	90.0	14.8	44.0	83.6	90.0	90.0	24.8	52.6	83.1	90.0	90.0
	8	4	55.7	18.4	31.3	58.4	86.0	90.0	9.0	29.4	62.0	75.1	90.0	21.0	30.5	66.2	82.5	90.0	28.6	32.9	71.1	90.0	90.0	35.4	48.7	74.6	90.0	90.0
	9	3	45.7	16.3	17.2	44.5	67.9	90.0	16.0	17.1	48.4	62.9	90.0	34.7	21.5	53.5	68.5	90.0	42.7	31.4	59.5	76.3	90.0	48.0	48.7	67.9	81.8	90.0
	10	2	37.0	23.1	4.0	30.6	53.9	81.9	27.9	11.8	35.5	46.9	79.3	48.5	29.6	42.1	54.7	88.6	56.8	39.3	49.6	62.9	90.0	62.0	50.8	63.5	70.6	90.0
	11	1	30.5	34.0	11.8	17.6	39.9	70.1	40.8	20.2	24.8	33.1	66.7	62.1	41.5	33.4	41.5	75.5	70.7	50.3	42.6	50.3	84.4	75.9	60.8	62.0	81.8	90.0
	12		28.0	45.9	25.8	8.0	25.8	58.1	53.8	33.1	20.0	33.1	53.8	62.1	38.5	30.0	41.5	62.1	70.7	40.0	40.0	50.3	70.7	90.0	70.6	62.0	70.6	90.0
June 21	5	7	82.1	49.2	72.5	90.0	90.0	90.0	41.4	71.1	90.0	90.0	90.0	34.5	70.2	90.0	90.0	90.0	29.2	70.0	90.0	90.0	90.0	27.6	71.7	90.0	90.0	90.0
	6	6	72.8	37.1	58.9	86.0	90.0	90.0	28.4	57.3	90.0	90.0	90.0	20.8	56.8	90.0	90.0	90.0	16.0	57.5	90.0	90.0	90.0	23.4	62.6	90.0	90.0	90.0
	7	5	63.0	25.3	45.2	72.0	90.0	90.0	15.7	43.7	76.3	90.0	90.0	7.1	44.1	80.0	90.0	90.0	15.8	46.2	85.0	90.0	90.0	27.6	55.7	85.7	90.0	90.0
	8	4	52.9	15.7	31.4	58.6	80.7	90.0	6.4	30.7	62.7	79.3	90.0	6.7	32.8	67.5	88.6	90.0	29.4	35.7	72.8	90.0	90.0	37.4	52.1	77.5	90.0	90.0
	9	3	43.1	14.6	17.6	44.8	57.3	79.9	14.0	18.2	49.6	63.6	90.0	20.5	25.2	55.5	82.2	90.0	43.5	32.6	61.3	77.2	90.0	49.6	52.1	71.1	81.8	90.0
	10	2	34.2	20.0	4.6	31.2	45.1	68.3	25.7	11.1	37.6	47.5	77.7	34.6	21.6	44.6	55.3	87.2	56.4	41.8	51.3	64.2	90.0	62.7	54.4	66.9	70.6	90.0
	11	1	27.3	32.2	13.0	19.6	61.1	56.4	39.5	22.2	27.6	41.8	65.3	47.8	31.9	36.5	55.3	74.4	70.1	52.2	44.8	52.2	85.3	76.3	63.6	65.4	81.8	90.0
	12		24.6	44.2	26.1	13.4	26.1	44.2	52.5	34.2	23.4	34.2	52.5	61.2	43.0	33.4	43.0	61.2	70.1	43.4	43.4	52.2	70.1	90.0	72.9	65.4	72.9	90.0

198

Solar Angle Of Incidence On Surfaces Tilted With Respect To The Horizontal

Date	Solar Time AM	PM	0°	38° E/W	38° SE/SW	38° S	38° SW/SE	38° W/E	48° E/W	48° SE/SW	48° S	48° SW/SE	48° W/E	58° E/W	58° SE/SW	58° S	58° SW/SE	58° W/E	68° E/W	68° SE/SW	68° S	68° SW/SE	68° W/E	90° E/W	90° SE/SW	90° S	90° SW/SE	90° W/E
July 21	5	7	84.3	50.6	73.1	90.0	90.0	90.0	42.4	71.1	90.0	90.0	90.0	35.0	69.8	90.0	90.0	90.0	28.9	69.0	90.0	90.0	90.0	25.3	69.8	90.0	90.0	90.0
	6	6	74.8	38.5	59.3	86.5	90.0	90.0	29.5	57.1	90.0	90.0	90.0	21.2	56.0	90.0	90.0	90.0	15.0	56.1	90.0	90.0	90.0	20.6	60.3	90.0	90.0	90.0
	7	5	64.9	27.1	45.3	72.6	81.4	90.0	17.2	43.2	76.0	89.0	90.0	7.7	42.8	79.8	90.0	90.0	4.4	44.3	83.8	90.0	90.0	25.3	53.1	83.1	90.0	90.0
	8	4	54.9	17.8	31.3	58.5	67.8	81.5	9.3	29.6	62.1	67.8	81.5	7.3	30.5	66.4	82.5	88.3	14.8	34.9	71.4	82.2	90.0	35.8	49.5	75.1	90.0	90.0
	9	3	45.5	15.7	18.5	44.5	53.9	67.8	15.7	17.5	48.6	53.9	67.8	20.8	22.1	53.8	67.5	75.3	28.7	31.9	59.9	76.4	90.0	48.6	53.8	68.5	90.0	90.0
	10	2	36.5	22.8	4.4	30.7	39.9	53.3	27.7	12.4	35.8	39.9	53.3	34.6	22.1	42.5	54.9	61.9	42.7	39.7	43.2	63.1	84.3	62.1	61.3	64.1	82.1	82.1
	11	1	29.9	33.7	12.0	17.9	25.8	57.7	40.6	20.5	25.3	25.8	57.7	48.3	30.0	34.0	41.7	57.7	56.7	50.6	40.6	63.1	70.6	76.0	77.5	64.1	71.0	71.0
	12	12	27.4	45.6	25.8	10.6	25.8	45.6	53.6	33.3	20.6	33.3	53.6	61.9	41.7	30.6	41.7	61.9	70.6	62.6	40.6	62.6	70.6	90.0	71.0	62.6	71.0	90.0
Aug 21	6	6	80.9	43.4	60.8	87.9	90.0	90.0	33.7	57.0	90.0	90.0	90.0	24.2	54.2	90.0	90.0	90.0	15.2	52.7	90.0	90.0	90.0	12.3	53.8	90.0	90.0	90.0
	7	5	70.9	33.1	46.5	73.0	90.0	90.0	23.0	42.4	75.4	87.2	90.0	13.1	39.9	77.9	90.0	90.0	3.9	39.5	80.5	90.0	90.0	19.3	45.6	87.4	90.0	90.0
	8	4	61.0	25.4	32.1	58.8	83.7	90.0	12.9	27.7	60.8	72.9	87.2	12.9	26.3	63.6	81.8	90.0	14.9	28.2	67.3	88.7	90.0	32.2	41.3	77.3	90.0	90.0
	9	3	51.6	23.9	18.0	44.2	69.5	86.3	23.8	13.3	46.3	58.3	72.4	23.8	15.1	50.2	67.5	82.1	29.2	21.8	54.8	71.9	90.0	46.3	41.3	68.2	88.7	90.0
	10	2	43.6	29.3	5.6	29.5	55.3	74.3	37.0	14.7	32.2	44.8	63.3	40.5	15.4	36.0	53.0	67.5	43.6	24.4	54.8	46.2	86.5	60.8	46.4	46.4	76.3	86.5
	11	1	37.8	38.9	12.6	14.9	40.9	62.2	50.7	17.3	19.3	34.1	61.1	50.7	23.8	26.8	42.3	61.1	58.0	34.1	35.3	46.2	63.1	75.4	55.6	55.6	62.2	62.2
	12	12	35.7	50.2	26.6	2.3	26.6	50.2	64.5	31.3	12.3	31.3	57.1	64.5	38.7	23.3	38.7	64.5	72.3	57.1	33.5	57.1	72.3	90.0	54.3	54.3	65.6	90.0
Sept 21	7	5	80.0	43.0	50.2	75.2	90.0	90.0	33.5	43.6	75.0	90.0	90.0	24.4	38.2	75.2	90.0	90.0	16.2	34.5	75.9	90.0	90.0	15.0	35.0	78.9	90.0	90.0
	8	4	70.5	37.2	36.6	60.5	87.2	90.0	29.8	29.0	60.0	77.2	90.0	21.8	23.2	60.5	81.9	90.0	21.8	20.5	62.0	86.0	90.0	30.0	28.9	68.2	90.0	90.0
	9	3	61.8	36.1	24.6	45.9	72.9	90.0	36.1	15.4	45.0	62.4	87.7	31.7	6.9	45.5	69.9	81.8	33.6	14.8	48.4	71.0	90.0	45.0	29.4	46.3	84.2	90.0
	10	2	54.6	40.1	17.4	31.5	58.3	89.1	45.0	8.6	30.0	47.6	76.1	43.0	6.9	31.5	51.8	82.9	45.0	28.1	35.5	71.9	89.9	65.0	46.4	44.1	71.1	89.9
	11	1	49.7	40.1	20.9	18.0	40.9	76.3	58.8	33.0	19.3	33.0	63.4	61.1	36.8	26.8	36.8	82.0	61.2	26.3	35.5	57.1	89.9	75.0	46.3	44.1	57.1	89.9
	12	12	48.0	5.2	31.7	10.0	31.7	58.2	69.5	37.1	12.3	37.1	63.4	69.2	36.8	10.0	36.8	69.2	75.5	20.0	32.3	42.4	75.5	90.0	58.3	54.3	58.3	90.0
Oct 21	7	5	88.0	52.3	54.8	77.4	90.0	90.0	43.2	46.7	76.3	90.0	90.0	34.5	39.2	73.6	90.0	90.0	32.9	34.4	72.5	90.0	90.0	18.2	26.9	71.9	90.0	90.0
	8	4	78.8	47.4	42.7	63.1	79.8	90.0	40.3	35.0	60.3	87.7	90.0	34.4	25.1	58.9	82.2	90.0	30.5	18.1	58.4	83.0	90.0	18.6	18.0	60.3	83.0	90.0
	9	3	70.7	46.5	32.8	49.2	76.4	87.7	42.5	22.8	46.0	62.4	90.0	40.1	12.2	44.2	62.5	81.9	39.8	13.6	44.3	70.4	90.0	45.0	19.4	41.0	79.3	90.0
	10	2	64.3	46.9	23.9	36.2	58.4	76.3	49.0	19.4	31.4	50.5	81.6	47.2	12.9	29.8	52.5	87.2	46.0	26.3	34.3	71.0	90.0	75.3	28.2	41.0	66.1	90.0
	11	1	60.0	52.1	30.1	20.5	38.5	76.3	58.9	29.1	18.6	37.1	81.6	61.1	19.8	26.3	34.3	82.5	61.3	17.3	34.3	55.7	90.0	75.3	35.9	31.1	55.6	90.0
	12	12	58.5	65.7	38.5	20.5	38.5	65.7	69.5	37.1	10.5	37.1	69.5	73.9	38.0	10.5	38.0	73.9	78.7	9.5	41.0	42.4	78.7	90.0	52.9	31.5	52.9	90.0
Nov 21	8	4	86.4	56.5	49.1	66.1	90.0	90.0	49.6	39.4	61.9	80.9	90.0	43.6	29.8	58.5	82.2	90.0	38.8	20.7	56.1	90.0	90.0	35.4	10.4	54.8	90.0	90.0
	9	3	78.8	55.8	40.9	53.3	79.8	90.0	48.2	30.9	51.6	67.3	90.0	46.4	21.0	44.4	68.1	90.0	46.9	11.1	53.3	83.7	90.0	48.3	11.4	43.9	87.7	90.0
	10	2	72.9	58.6	37.0	42.0	67.1	82.5	56.9	28.2	35.4	54.2	86.7	56.9	20.5	30.5	54.3	90.0	56.9	15.4	28.1	69.7	90.0	61.9	22.9	33.7	75.2	90.0
	11	1	69.1	64.5	33.3	29.8	55.4	82.5	65.2	32.8	24.7	34.2	86.7	68.9	28.9	17.5	34.3	90.0	68.9	27.8	24.1	55.6	81.9	75.9	35.9	25.6	62.2	90.0
	12	12	67.8	72.7	45.5	29.8	45.5	72.7	75.4	42.3	19.8	42.3	75.4	78.4	40.9	9.8	40.9	78.4	81.9	0.2	41.5	41.5	81.9	90.0	49.1	22.2	49.1	90.0
Dec 21	9	3	82.0	59.4	44.2	55.3	81.2	90.0	54.9	34.2	49.6	81.7	90.0	51.4	24.3	44.9	82.4	90.0	49.2	14.6	41.8	83.3	90.0	49.6	8.9	41.6	86.0	90.0
	10	2	74.7	62.1	40.7	44.5	69.0	90.0	60.1	31.9	37.6	68.5	90.0	59.1	24.0	31.5	68.7	90.0	62.7	18.1	28.0	71.0	90.0	62.7	21.5	31.1	66.0	90.0
	11	1	72.7	67.6	42.3	36.5	57.8	90.0	68.0	35.9	27.6	55.9	88.7	69.0	31.3	19.6	34.3	90.0	70.6	29.2	14.3	55.3	90.0	76.3	34.7	22.4	60.9	90.0
	12	12	71.4	75.5	48.4	33.4	48.4	75.5	77.7	44.7	23.4	44.7	77.7	80.3	42.5	13.5	42.5	80.3	83.2	1.5	42.2	42.2	83.2	90.0	47.9	18.5	47.9	90.0

aBased on data in Table 1, p. 387, *1972 ASHRAE Handbook of Fundamentals*; 0% ground reflectance, 1.0 clearness factor.

Table A.13. Solar Angle of Incidence on Inclined Surfaces at 56° North Latitude

Solar Angle of Incidence On Surfaces Tilted With Respect To The Horizontal

Date	AM	PM	0°	46° E/W	46° SE/SW	46° S	46° SW/SE	46° W/E	56° E/W	56° SE/SW	56° S	56° SW/SE	56° W/E	66° E/W	66° SE/SW	66° S	66° SW/SE	66° W/E	76° E/W	76° SE/SW	76° S	76° SW/SE	76° W/E	90° E/W	90° SE/SW	90° S	90° SW/SE	90° W/E
Jan. 21	9	3	85.0	57.4	39.0	53.5	84.2	90.0	53.1	29.1	48.4	84.6	90.0	50.0	19.2	44.5	85.1	90.0	48.2	9.5	42.2	85.7	90.0	48.4	6.0	42.1	86.9	90.0
	10	2	80.1	62.8	37.0	42.1	73.5	90.0	61.0	28.5	35.5	70.8	90.0	60.1	21.2	30.6	71.0	90.0	60.2	16.7	28.2	71.7	90.0	62.0	19.2	30.0	73.7	90.0
	11	1	77.1	70.7	40.6	33.4	59.2	90.0	70.9	34.8	24.8	57.6	90.0	71.2	30.5	17.6	57.1	90.0	73.1	29.7	14.1	57.7	90.0	75.9	32.9	19.3	60.3	90.0
	12		76.0	80.3	48.6	30.0	48.6	80.3	82.2	44.2	20.0	43.2	82.2	84.4	43.5	10.0	43.5	84.4	86.6	43.6	0.0	43.6	86.6	90.0	46.7	14.0	46.7	90.0
Feb. 21	8	4	82.5	45.2	38.5	62.9	90.0	90.0	38.7	29.6	60.5	90.0	90.0	33.6	21.4	59.0	90.0	90.0	30.8	15.5	58.5	90.0	90.0	31.5	16.2	59.6	90.0	90.0
	9	3	75.8	47.8	29.8	49.0	80.8	90.0	44.4	19.8	45.9	82.9	90.0	42.6	9.8	44.3	85.1	90.0	42.7	0.9	44.5	87.5	90.0	45.9	14.2	47.6	90.0	90.0
	10	2	70.0	54.7	27.0	35.9	67.1	90.0	53.5	18.9	31.5	68.3	90.0	54.2	13.4	29.5	70.3	90.0	56.1	14.0	30.6	72.9	90.0	60.5	23.5	36.5	77.3	90.0
	11	1	67.2	63.1	25.9	25.0	53.8	85.1	63.6	21.3	18.0	54.1	89.7	65.3	26.5	17.6	55.6	89.7	68.4	28.5	17.6	58.2	90.0	75.2	36.2	27.7	63.5	90.0
	12		66.0	73.6	41.6	20.0	41.6	73.6	76.9	40.3	10.0	40.3	76.9	80.5	40.9	0.9	40.9	80.5	84.4	43.5	10.0	43.5	84.4	90.0	49.8	24.0	49.8	90.0
Mar. 21	7	5	81.7	37.3	45.5	75.2	90.0	90.0	28.1	39.4	75.0	90.0	90.0	19.7	34.8	75.2	90.0	90.0	13.5	32.3	75.9	90.0	90.0	15.0	33.4	77.6	90.0	90.0
	8	4	73.8	38.4	32.3	60.5	90.0	90.0	29.0	24.9	60.0	90.0	90.0	25.2	19.8	60.5	90.0	90.0	24.8	18.9	62.0	89.3	90.0	30.0	25.1	65.5	90.0	90.0
	9	3	66.7	45.9	21.7	45.5	77.1	90.0	36.3	11.7	45.0	81.4	90.0	36.2	10.3	45.3	85.3	90.0	36.1	10.7	48.4	74.5	90.0	45.0	23.9	54.1	89.3	90.0
	10	2	61.0	55.8	17.1	31.5	63.4	90.0	46.7	10.0	30.0	66.5	90.0	49.2	10.3	31.5	70.3	90.0	49.2	17.7	35.5	59.8	90.0	60.0	30.5	44.5	74.5	90.0
	11	1	57.3	65.5	23.9	18.0	49.4	79.1	58.9	22.6	15.0	51.7	85.0	62.9	25.3	18.0	55.3	89.7	63.0	30.9	24.8	45.2	82.2	75.0	41.5	36.8	59.4	90.0
	12		56.0	67.1	35.9	10.0	35.9	67.1	71.8	37.0	0.0	37.0	71.8	76.9	40.3	10.0	40.3	76.9	82.2	45.2	20.0	45.2	82.2	90.0	54.1	34.0	54.1	90.0
Apr. 21	5	7	88.6	45.7	70.5	90.0	90.0	90.0	37.0	67.7	90.0	90.0	90.0	29.0	65.6	90.0	90.0	90.0	22.5	64.3	90.0	90.0	90.0	18.9	63.8	90.0	90.0	90.0
	6	6	80.4	34.9	56.2	88.0	90.0	90.0	25.1	53.0	90.0	90.0	90.0	15.7	51.1	90.0	90.0	90.0	6.9	50.6	90.0	90.0	90.0	11.6	52.2	90.0	90.0	90.0
	7	5	72.0	26.5	41.8	73.5	90.0	90.0	16.9	38.1	75.3	90.0	90.0	8.2	36.9	77.6	90.0	90.0	6.9	37.1	80.2	90.0	90.0	18.9	42.4	84.0	90.0	90.0
	8	4	63.9	23.6	27.5	58.9	88.6	90.0	18.3	23.7	60.7	80.0	90.0	17.4	23.5	63.4	85.6	90.0	21.6	26.3	67.0	90.0	90.0	32.0	36.1	72.9	90.0	90.0
	9	3	56.4	28.0	13.6	44.5	74.5	90.0	27.9	9.4	46.2	65.3	90.0	30.9	13.8	49.7	71.0	90.0	36.3	23.3	54.4	77.0	90.0	46.2	35.3	62.5	90.0	90.0
	10	2	50.1	37.1	5.7	29.5	60.2	84.7	40.1	7.3	32.0	50.9	90.0	44.9	16.5	36.8	56.4	90.0	50.9	26.3	43.2	62.9	90.0	60.7	40.2	53.8	79.9	90.0
	11	1	47.2	47.5	17.4	14.8	47.5	72.5	54.5	21.3	18.9	35.9	79.7	59.0	28.3	24.3	42.0	87.1	65.5	36.6	36.6	49.2	90.0	75.3	49.1	47.0	65.1	90.0
	12		44.4	60.2	31.5	1.6	31.5	60.2	66.5	35.9	11.6	35.9	66.5	73.1	42.0	21.6	42.0	73.1	80.0	49.2	31.6	49.2	80.0	90.0	60.3	45.6	60.3	90.0
May 21	4	8	88.8	53.0	82.4	90.0	90.0	90.0	45.6	81.2	90.0	90.0	90.0	41.2	80.9	90.0	90.0	90.0	37.9	81.4	90.0	90.0	90.0	35.5	80.5	90.0	90.0	90.0
	5	7	81.5	40.9	68.6	90.0	90.0	90.0	33.6	67.4	90.0	90.0	90.0	27.2	66.9	90.0	90.0	90.0	24.8	68.4	90.0	90.0	90.0	24.8	67.1	90.0	90.0	90.0
	6	6	73.5	29.3	54.7	86.6	90.0	90.0	22.6	53.3	89.0	90.0	90.0	13.0	53.2	90.0	90.0	90.0	14.1	56.2	90.0	90.0	90.0	13.6	54.4	90.0	90.0	90.0
	7	5	65.2	19.2	40.7	72.6	90.0	90.0	11.4	39.4	74.6	86.0	90.0	1.0	40.2	78.6	86.0	90.0	13.6	45.3	78.6	90.0	90.0	16.0	43.0	78.6	90.0	90.0
	8	4	56.9	15.2	26.6	58.5	86.2	90.0	15.5	23.5	62.0	72.0	90.0	15.1	28.8	66.2	72.0	90.0	22.8	34.3	71.6	86.2	90.0	29.8	33.6	68.9	90.0	90.0
	9	3	49.1	21.1	12.6	44.5	72.5	90.0	26.3	13.6	47.4	58.2	88.7	29.1	21.9	53.5	58.2	90.0	36.6	34.3	61.1	72.5	90.0	42.8	34.3	61.1	78.6	90.0
	10	2	42.4	31.7	3.7	30.6	58.7	79.9	36.6	3.7	34.9	44.9	83.2	43.1	23.6	42.1	44.9	84.1	50.8	36.8	52.4	59.1	90.0	56.5	47.0	55.9	67.2	90.0
	11	1	37.7	43.6	16.5	17.6	44.7	68.0	49.6	25.7	27.6	51.5	74.6	57.0	32.7	33.4	44.9	70.8	64.5	44.4	45.8	46.4	83.2	69.9	58.0	54.0	67.2	78.2
	12		36.0	55.8	30.8	11.4	30.8	55.8	63.1	38.1	21.6	38.1	63.1	70.8	44.9	33.4	44.9	70.8	78.2	55.0	43.4	55.0	69.9	90.0	67.6	57.4	67.6	90.0
June 21	4	8	85.8	51.5	81.5	90.0	90.0	90.0	45.6	81.2	90.0	90.0	90.0	40.9	81.2	90.0	90.0	90.0	37.4	82.2	90.0	90.0	90.0	37.4	82.2	90.0	90.0	90.0
	5	7	78.6	39.3	68.0	90.0	90.0	90.0	32.3	67.6	90.0	90.0	90.0	27.2	67.6	90.0	90.0	90.0	24.8	68.9	90.0	90.0	90.0	27.4	70.7	90.0	90.0	90.0
	6	6	70.7	27.2	54.4	86.0	90.0	90.0	20.9	53.7	89.2	90.0	90.0	13.9	54.0	90.0	90.0	90.0	13.6	56.6	88.0	90.0	90.0	23.4	60.6	88.0	90.0	90.0
	7	5	62.4	16.5	40.9	72.6	90.0	90.0	13.4	40.9	72.3	85.9	90.0	3.9	41.4	80.0	86.2	90.0	13.9	45.7	80.1	90.0	90.0	27.4	52.7	78.6	90.0	90.0
	8	4	54.1	13.7	26.9	58.6	85.2	90.0	17.3	25.3	62.6	72.5	90.0	13.3	30.0	67.3	72.5	90.0	23.4	36.0	71.6	86.3	90.0	37.4	47.0	71.6	90.0	90.0
	9	3	46.2	18.6	13.7	44.9	71.8	89.2	29.0	19.8	49.6	59.1	87.0	29.0	25.3	55.3	59.1	87.0	37.4	34.0	61.7	72.5	90.0	49.6	50.7	64.1	78.2	90.0
	10	2	39.3	29.8	6.8	31.6	58.2	78.8	37.4	6.8	34.9	51.5	83.2	42.8	26.8	44.5	51.5	83.2	51.3	36.8	52.4	59.1	90.0	62.7	58.0	59.2	67.2	90.0
	11	1	34.4	41.9	17.3	19.6	44.5	66.3	50.8	25.9	34.9	46.3	59.1	56.5	34.9	36.5	46.3	69.9	64.5	45.8	45.8	47.0	83.2	76.3	67.6	58.0	55.0	78.2
	12		32.6	54.2	30.8	13.1	30.8	54.2	64.5	46.3	33.4	46.3	69.9	69.9	46.3	33.4	46.3	69.9	78.2	55.0	43.4	55.0	69.9	90.0	67.6	57.4	67.6	90.0

Solar Angle Of Incidence On Surfaces Tilted With Respect To The Horizontal

Date	Solar Time AM	Solar Time PM	0°	46° E/W	46° SE/SW	46° S	46° SW/SE	46° W/E	56° E/W	56° SE/SW	56° S	56° SW/SE	56° W/E	66° E/W	66° SE/SW	66° S	66° SW/SE	66° W/E	76° E/W	76° SE/SW	76° S	76° SW/SE	76° W/E	90° E/W	90° SE/SW	90° S	90° SW/SE	90° W/E
July 21	4	8	88.-	52.8	82.2	90.0	90.0	90.0	46.5	81.4	90.0	90.0	90.0	41.2	80.9	90.0	90.0	90.0	37.5	80.7	90.0	90.0	90.0	35.8	80.8	90.0	90.0	90.0
	5	7	81.0	40.6	68.5	87.9	90.0	90.0	33.2	67.4	90.0	90.0	90.0	27.2	67.0	90.0	90.0	90.0	23.8	67.3	90.0	90.0	90.0	25.3	69.0	90.0	90.0	90.0
	6	6	73.0	28.8	54.6	73.4	86.5	90.0	20.1	53.4	76.0	90.0	90.0	13.2	53.4	77.8	90.0	90.0	11.8	54.7	80.5	90.0	90.0	20.6	58.5	84.8	90.0	90.0
	7	5	64.7	18.7	40.6	72.6	86.0	90.0	8.7	39.5	76.0	90.0	90.0	1.3	40.5	79.8	90.0	90.0	11.3	43.4	71.8	90.0	90.0	25.3	50.0	63.0	89.8	90.0
	8	4	56.4	16.6	24.2	58.5	72.6	90.0	11.0	14.2	62.1	65.1	90.0	15.1	24.2	66.4	72.1	90.0	23.0	34.2	71.8	79.3	90.0	35.3	48.2	69.0	89.8	90.0
	9	3	48.4	26.6	4.2	44.5	58.6	90.0	26.4	14.2	50.6	65.1	90.0	26.6	24.2	56.9	72.1	90.0	36.6	34.2	63.2	79.3	90.0	48.2	55.8	61.6	89.7	90.0
	10	2	41.8	31.1	16.6	30.7	44.6	79.6	36.4	24.2	35.8	51.1	84.1	43.1	34.0	37.2	58.3	84.1	50.6	42.4	43.2	66.6	79.3	62.1	44.2	54.6	77.3	89.7
	11	1	37.1	43.1	17.9	44.6	30.6	67.7	49.7	14.2	25.3	37.3	75.8	56.9	33.1	34.0	43.2	75.8	64.7	53.5	35.5	53.1	90.0	76.0	55.8	54.6	77.3	89.7
	12	12	35.4	55.5	30.6	10.6	30.6	55.5	62.9	37.3	20.6	37.3	62.9	70.6	45.1	30.6	45.1	70.6	78.6	53.5	40.6	53.5	78.6	90.0	65.8	54.6	65.8	90.0
Aug. 21	5	7	88.0	45.3	70.3	90.0	90.0	90.0	36.7	67.7	90.0	90.0	90.0	28.8	65.7	90.0	90.0	90.0	22.4	64.5	90.0	90.0	90.0	19.3	64.2	90.0	90.0	90.0
	6	6	79.8	34.4	56.0	87.9	90.0	90.0	24.7	54.4	90.0	90.0	90.0	15.7	51.2	90.0	90.0	90.0	7.8	50.9	90.0	90.0	90.0	12.3	52.7	90.0	90.0	90.0
	7	5	71.5	25.9	41.7	73.4	86.0	90.0	16.2	38.4	75.4	90.0	90.0	7.5	37.1	77.8	90.0	90.0	6.9	38.1	80.5	90.0	90.0	19.3	43.0	84.8	90.0	90.0
	8	4	63.3	22.9	27.3	58.8	74.3	90.0	17.6	9.7	46.3	58.4	90.0	17.3	23.9	63.6	72.0	90.0	21.6	33.7	54.8	79.0	90.0	32.2	36.8	73.4	90.0	90.0
	9	3	55.7	27.4	13.1	44.2	58.4	84.0	27.4	7.6	32.2	46.5	79.9	30.7	14.5	37.2	53.7	85.7	30.9	22.9	40.8	70.0	90.0	46.6	36.0	63.0	86.2	90.0
	10	2	49.2	36.6	5.1	29.5	44.6	72.1	39.8	21.4	28.6	50.6	65.5	44.7	17.1	26.8	37.0	86.8	37.0	37.0	40.6	63.1	90.0	54.4	49.6	54.8	86.2	73.1
	11	1	45.2	47.8	17.2	14.9	30.6	75.9	52.9	21.4	12.3	37.0	79.4	58.8	28.6	22.3	42.2	86.8	28.8	49.5	32.3	49.5	79.9	75.4	49.6	48.4	73.1	60.8
	12	12	43.7	59.9	31.4	2.3	31.4	59.9	67.1	36.0	12.3	36.0	67.1	72.9	42.2	22.3	42.2	72.9	79.9	49.5	35.5	49.5	79.9	90.0	60.8	46.3	60.8	90.0
Sept. 21	7	5	81.7	37.3	45.5	75.2	90.0	90.0	28.1	39.4	75.0	90.0	90.0	19.7	34.8	75.2	90.0	90.0	13.5	32.3	75.9	90.0	90.0	15.0	33.4	77.6	90.0	90.0
	8	4	73.8	32.2	32.3	60.5	77.0	90.0	29.0	24.9	60.0	90.0	90.0	25.2	19.8	60.5	90.0	90.0	24.8	18.9	62.0	90.0	90.0	30.0	25.1	65.5	73.8	90.0
	9	3	66.7	38.4	21.2	45.9	77.2	90.0	36.2	11.7	45.0	81.4	90.0	36.2	4.9	45.9	85.3	90.0	38.6	11.7	48.4	89.3	90.0	45.0	23.9	54.1	90.0	90.0
	10	2	61.0	15.9	17.1	18.0	63.4	90.0	46.7	10.0	15.0	51.7	85.0	49.2	10.3	31.5	70.3	90.0	53.0	10.6	24.8	59.8	90.0	60.0	30.5	36.8	81.0	90.0
	11	1	57.3	5.8	23.9	14.9	49.4	84.1	58.9	22.6	0.0	37.0	90.0	62.9	25.3	18.0	55.3	90.0	67.6	28.8	35.5	45.2	90.0	75.0	41.5	34.0	67.5	90.0
	12	12	56.0	67.1	35.9	2.3	35.1	73.9	71.8	37.0	12.3	37.0	71.8	76.9	40.3	22.3	45.2	76.9	90.0	45.9	35.9	45.2	82.2	90.0	54.1	34.0	54.1	90.0
Oct. 21	8	4	82.9	45.7	38.9	63.1	90.0	90.0	39.2	29.9	60.6	90.0	90.0	34.1	21.6	58.9	90.0	90.0	31.1	15.5	58.4	90.0	90.0	31.6	15.8	59.4	90.0	90.0
	9	3	76.2	48.3	30.3	49.2	81.0	90.0	44.8	20.3	46.0	82.9	90.0	43.0	10.3	44.2	85.1	90.0	42.9	0.7	44.3	87.4	90.0	46.0	13.8	47.3	90.0	90.0
	10	2	71.0	54.6	27.5	42.6	67.3	85.6	53.9	19.3	34.5	68.5	90.0	54.5	10.3	29.5	73.0	90.0	56.2	14.0	30.4	72.8	90.0	60.6	23.2	36.1	77.1	90.0
	11	1	67.5	63.4	27.5	20.5	54.1	67.3	64.9	18.5	10.5	40.5	67.0	67.5	26.1	0.5	55.6	90.0	70.2	28.5	17.3	43.5	84.5	75.3	36.6	27.2	63.6	90.0
	12	12	66.5	71.9	42.0	20.5	42.0	73.9	77.1	40.5	10.5	40.5	77.1	80.7	42.6	0.5	41.0	80.7	84.5	43.5	9.5	43.5	84.5	90.0	49.6	23.5	49.6	90.0
Nov. 21	9	3	84.8	57.2	38.9	53.4	84.2	90.0	52.9	28.9	48.3	84.5	90.0	49.8	19.0	44.4	85.1	90.0	48.1	9.3	42.2	85.8	90.0	48.3	6.1	42.2	86.9	90.0
	10	2	79.9	62.6	36.8	42.3	51.2	90.0	60.8	28.3	35.4	70.8	90.0	59.9	11.0	30.5	51.0	90.0	60.1	16.6	28.2	71.8	90.0	61.9	19.2	30.1	73.8	90.0
	11	1	76.9	76.5	40.4	33.3	59.1	90.0	70.8	24.7	24.7	57.5	90.0	84.3	30.8	17.5	57.0	90.0	86.6	29.7	14.1	43.6	86.6	90.0	33.0	19.5	46.7	90.0
	12	12	75.8	80.2	48.4	29.8	48.4	80.2	82.1	45.1	19.8	45.1	82.1	84.3	43.4	9.8	43.4	84.3	86.6	43.6	0.2	43.6	86.6	90.0	46.7	14.2	46.7	90.0
Dec. 21	9	3	88.1	60.7	43.3	55.3	85.4	90.0	56.2	32.4	49.6	85.2	90.0	52.7	22.5	44.9	85.1	90.0	50.4	12.9	41.8	85.2	90.0	49.6	4.9	40.5	85.5	90.0
	10	2	83.4	65.8	40.7	36.5	72.9	90.0	61.6	31.9	28.2	59.4	90.0	62.5	24.2	31.6	51.4	90.0	61.8	18.7	28.0	51.7	90.0	62.8	18.7	28.2	73.4	90.0
	11	1	80.5	73.6	40.4	33.4	61.2	90.0	73.8	31.5	23.4	58.0	90.0	85.7	33.5	13.5	44.8	90.0	87.5	30.7	3.4	43.6	87.5	90.0	32.9	16.8	59.4	90.0
	12	12	79.4	82.7	51.2	33.4	51.2	82.7	84.1	47.3	23.4	47.3	84.1	85.7	44.8	13.5	44.8	85.7	87.5	44.0	3.4	44.0	87.5	90.0	46.0	10.5	46.0	90.0

aBased on data in Table 1, p. 387, 1972 *ASHRAE Handbook of Fundamentals*; 0% ground reflectance, 1.0 clearness factor.

Table A.14. Solar Angle of Incidence on Inclined Surfaces at 64° North Latitude

Solar Angle Of Incidence On Surfaces Tilted With Respect To The Horizontal

Date	Solar Time AM/PM	D°	54° E/W	54° SE/SW	54° S/S	54° SW/SE	54° W/E	64° E/W	64° SE/SW	64° S/S	64° SW/SE	64° W/E	74° E/W	74° SE/SW	74° S/S	74° SW/SE	74° W/E	84° E/W	84° SE/SW	84° S/S	84° SW/SE	84° W/E	90° E/W	90° SE/SW	90° S/S	90° SW/SE	90° W/E
Jan. 21	10/2	87.2	65.9	36.7	42.1	74.7	90.0	63.6	28.3	35.5	73.6	90.0	62.3	21.2	30.6	72.9	90.0	61.8	17.2	28.2	72.9	90.0	62.0	17.2	28.2	73.1	90.0
	11/1	84.8	75.5	41.9	33.4	62.2	90.0	75.0	36.1	24.8	60.1	90.0	75.0	32.1	17.6	58.9	90.0	75.4	30.7	14.1	58.8	90.0	75.9	31.3	15.0	59.3	90.0
	12	84.0	86.5	50.9	30.0	50.9	86.5	87.4	47.3	20.0	47.3	87.4	88.3	45.2	10.0	45.2	88.3	89.4	44.7	90.0	44.7	89.4	90.0	45.3	6.0	45.3	90.0
Feb. 21	8/4	86.6	43.5	34.9	62.9	90.0	90.0	37.6	26.1	60.5	86.0	90.0	33.2	18.4	59.0	90.0	90.0	31.3	13.9	58.5	90.0	90.0	31.5	14.1	58.8	90.0	90.0
	9/3	81.4	57.5	27.4	48.8	84.8	90.0	56.3	17.4	45.9	86.5	90.0	44.7	7.4	44.3	87.7	90.0	44.9	2.6	44.5	88.9	90.0	45.9	8.6	45.4	89.8	90.0
	10/2	77.4	68.9	33.7	35.0	70.8	90.0	59.9	29.8	18.0	57.0	90.0	71.5	28.6	14.8	57.9	90.0	59.2	15.9	30.6	59.4	90.0	60.5	19.3	32.6	75.7	90.0
	11/1	74.9	80.7	44.6	25.0	57.3	90.0	69.9	43.0	10.0	43.0	90.0	85.6	43.2	10.6	43.2	85.6	77.7	30.1	12.6	45.2	88.3	75.2	43.7	14.1	68.9	88.7
	12	74.0	83.1	50.9	20.0	44.6	80.7	80.3	43.2	10.0	43.0	83.1	85.6	43.2	10.0	57.7	85.6	88.3	45.2	10.0	45.2	88.3	90.0	47.2	16.0	47.2	90.0
Mar. 21	7/5	83.5	32.0	41.2	75.2	90.0	90.0	23.4	35.8	75.0	90.0	90.0	16.3	32.2	75.2	90.0	90.0	13.5	31.3	75.9	90.0	90.0	15.0	32.1	76.5	90.0	90.0
	8/4	77.3	34.0	28.2	60.5	90.0	90.0	29.5	21.2	60.0	84.8	90.0	26.8	11.3	60.9	90.0	90.0	26.7	12.4	69.6	90.0	90.0	30.0	18.5	63.3	90.0	90.0
	9/3	71.9	41.0	18.1	45.9	81.9	90.0	39.5	8.4	45.0	84.8	90.0	42.6	3.6	45.9	87.9	90.0	42.6	20.2	48.4	90.0	90.0	45.0	25.3	48.3	78.6	90.0
	10/2	67.7	51.1	17.4	31.5	67.5	87.7	52.0	11.8	30.0	69.9	90.0	57.5	13.2	31.5	72.9	90.0	57.5	33.2	35.5	76.4	90.0	60.0	37.2	38.9	64.5	90.0
	11/1	64.9	62.7	26.7	18.0	53.3	77.9	52.0	28.0	15.0	55.0	90.0	72.4	18.0	18.0	57.9	90.0	72.4	33.2	24.8	61.8	90.0	75.0	37.2	29.8	78.6	90.0
	12	64.0	75.1	39.5	10.0	39.5	75.1	78.9	43.0	90.0	40.2	78.9	87.4	43.0	10.0	57.9	83.1	87.4	47.3	20.0	47.3	87.4	90.0	50.5	26.0	50.5	90.0
Apr. 21	5/7	86.0	36.3	66.3	90.0	90.0	90.0	28.2	64.5	90.0	90.0	90.0	21.8	63.4	90.0	90.0	90.0	18.5	63.2	90.0	90.0	90.0	18.9	63.5	90.0	90.0	90.0
	6/6	73.0	26.0	52.0	98.0	90.0	90.0	16.3	49.8	90.0	90.0	90.0	7.5	49.0	90.0	90.0	90.0	6.7	49.8	90.0	90.0	90.0	11.6	50.9	82.0	90.0	90.0
	7/5	73.0	20.4	36.7	73.5	90.0	90.0	11.9	35.1	75.3	90.0	90.0	8.1	35.1	77.6	90.0	90.0	13.7	37.4	80.2	90.0	90.0	18.9	39.8	87.0	90.0	90.0
	8/4	61.0	23.1	9.4	44.2	78.8	90.0	4.2	10.6	60.7	90.0	90.0	22.4	22.3	63.4	90.0	90.0	27.7	27.7	67.0	90.0	90.0	32.0	31.9	69.4	90.0	90.0
	9/3	56.5	32.3	7.8	29.5	64.5	89.4	17.0	10.9	32.0	68.9	90.0	22.0	19.3	36.8	73.9	90.0	42.3	24.8	34.6	79.2	90.0	46.2	29.8	57.7	69.4	90.0
	10/2	53.5	43.9	21.4	14.8	50.1	81.7	47.0	25.0	32.0	64.5	90.0	51.5	26.2	46.8	69.2	90.0	57.0	32.4	54.9	65.2	90.0	57.7	29.8	57.7	82.5	90.0
	11/1	52.4	56.3	35.7	1.6	35.7	69.0	60.7	39.6	11.6	39.6	88.1	65.9	45.0	21.6	45.0	90.0	71.7	39.0	31.6	51.6	86.3	73.3	44.1	37.6	68.9	90.0
May 21	4/8	84.2	44.1	78.6	90.0	90.0	90.0	39.1	78.6	90.0	90.0	90.0	35.9	78.9	90.0	90.0	90.0	34.9	79.6	90.0	90.0	90.0	35.5	80.2	90.0	90.0	90.0
	5/7	78.4	31.1	64.8	90.0	90.0	90.0	25.3	64.5	90.0	90.0	90.0	21.9	65.1	90.0	90.0	90.0	22.5	66.4	90.0	90.0	90.0	24.8	67.6	90.0	90.0	90.0
	6/6	72.1	19.8	50.8	86.6	90.0	90.0	11.7	50.5	90.0	90.0	90.0	8.9	51.6	90.0	90.0	90.0	14.8	54.0	83.6	90.0	90.0	20.0	56.1	86.1	90.0	90.0
	7/5	65.5	12.1	36.7	72.6	90.0	90.0	4.2	36.7	75.9	90.0	90.0	9.4	38.9	79.6	90.0	90.0	14.8	43.1	71.1	90.0	90.0	24.8	46.3	58.3	90.0	90.0
	8/4	59.2	16.2	24.9	58.5	90.0	90.0	17.0	13.7	62.4	90.0	90.0	22.5	28.4	66.2	90.0	90.0	30.3	35.2	75.1	90.0	90.0	35.5	39.8	79.8	90.0	90.0
	9/3	53.2	27.2	7.3	30.6	63.0	89.4	44.5	4.2	48.4	82.9	90.0	36.5	23.0	53.1	73.2	90.0	43.7	32.4	57.4	68.0	90.0	48.4	38.2	63.4	85.5	90.0
	10/2	48.4	39.6	20.8	17.6	49.0	77.4	58.1	27.5	34.8	54.7	84.8	57.6	35.7	42.6	61.2	90.0	57.6	36.5	42.6	68.2	90.0	52.0	32.0	54.4	72.6	90.0
	11/1	45.1	52.3	34.9	13.5	34.9	65.0	71.6	40.8	26.0	40.8	71.6	64.6	47.9	33.0	47.9	78.6	62.0	44.5	40.0	55.7	85.7	52.9	32.0	54.4	60.6	90.0
June 21	3/9	85.8	55.4	90.0	90.0	90.0	90.0	52.0	90.0	90.0	90.0	90.0	49.9	90.0	90.0	90.0	90.0	49.2	90.0	90.0	90.0	90.0	49.6	90.0	90.0	90.0	90.0
	4/8	81.0	42.7	77.8	90.0	90.0	90.0	38.5	78.4	90.0	90.0	90.0	36.2	79.4	90.0	90.0	90.0	36.2	80.6	90.0	90.0	90.0	37.4	81.6	90.0	90.0	90.0
	5/7	75.3	30.0	64.3	86.0	90.0	90.0	24.8	64.7	90.0	90.0	90.0	22.8	65.9	90.0	90.0	90.0	26.7	67.9	90.0	90.0	90.0	27.6	65.9	87.0	90.0	90.0
	6/6	69.0	17.7	50.6	72.6	90.0	90.0	11.1	51.0	76.3	90.0	90.0	18.9	52.9	90.0	90.0	90.0	18.9	56.0	85.0	90.0	90.0	23.4	58.3	90.0	90.0	90.0
	7/5	56.0	13.8	36.8	58.6	90.0	90.0	2.7	40.9	80.5	90.0	90.0	18.2	45.7	67.3	90.0	90.0	31.9	45.7	72.8	90.0	90.0	27.6	58.3	76.2	90.0	90.0
	8/4	50.1	25.6	23.0	44.9	76.1	90.0	17.6	17.6	49.6	82.0	90.0	31.6	31.8	67.5	90.0	90.0	31.9	38.3	72.8	90.0	90.0	37.4	49.1	87.8	90.0	90.0
	9/3	45.1	38.2	9.6	30.6	62.5	89.4	36.5	19.3	37.4	68.8	90.0	36.8	23.0	53.3	75.5	90.0	36.8	33.8	72.8	90.0	90.0	41.6	43.1	65.9	90.0	90.0
	10/2	41.7	50.9	31.6	19.6	48.8	75.7	57.3	29.2	37.4	55.1	90.0	57.5	37.8	45.1	62.2	90.0	57.1	39.1	52.4	70.5	90.0	62.7	41.6	57.3	88.7	90.0
	11/1	40.6	63.5	35.1	13.5	35.1	63.5	70.5	41.7	23.4	41.7	70.5	85.4	49.4	33.4	49.4	77.9	85.4	43.4	40.0	57.5	85.4	76.3	52.6	49.4	74.2	90.0

202

Solar Angle Of Incidence On Surfaces Tilted With Respect To The Horizontal

Date	Solar Time AM	PM	0°	54° E/W	54° SE/SW	54° S	54° SW/SE	54° W/E	64° E/W	64° SE/SW	64° S	64° SW/SE	64° W/E	74° E/W	74° SE/SW	74° S	74° SW/SE	74° W/E	84° E/W	84° SE/SW	84° S	84° SW/SE	84° W/E	90° E/W	90° SE/SW	90° S	90° SW/SE	90° W/E
July 21	4	8	83.6	43.9	78.5	90.0	90.0	90.0	39.0	78.5	90.0	90.0	90.0	35.9	79.0	90.0	90.0	90.0	35.1	79.8	90.0	90.0	90.0	35.8	80.4	90.0	90.0	90.0
	5	7	77.9	31.2	64.7	90.0	90.0	90.0	25.2	64.5	90.0	90.0	90.0	22.0	65.2	90.0	90.0	90.0	22.9	66.7	90.0	90.0	90.0	25.3	67.9	90.0	90.0	90.0
	6	6	71.6	19.4	50.7	86.5	90.0	90.0	11.5	50.5	90.0	90.0	90.0	9.3	51.8	90.0	87.9	90.0	15.4	54.3	90.0	90.0	90.0	20.6	56.4	86.4	90.0	90.0
	7	5	65.0	11.5	36.7	72.6	90.0	90.0	13.8	36.9	76.0	90.0	90.0	22.6	39.3	78.9	90.0	90.0	19.4	43.7	83.7	90.0	90.0	25.3	46.8	74.5	90.0	90.0
	8	4	58.5	13.8	23.0	58.5	90.0	90.0	30.6	14.9	63.9	90.0	90.0	36.5	23.6	68.8	88.8	90.0	43.8	33.0	71.4	81.7	90.0	35.8	40.4	63.9	90.0	90.0
	9	3	52.7	26.9	9.2	44.5	76.7	89.1	44.4	16.9	48.6	82.8	90.0	50.6	26.7	53.8	75.1	87.8	57.6	36.6	59.1	68.5	90.0	48.6	38.8	54.9	85.7	90.0
	10	2	47.8	39.3	7.7	30.7	62.9	77.1	58.0	27.8	35.8	68.9	84.6	64.6	39.5	34.0	61.4	74.2	71.6	44.9	43.2	68.5	90.0	62.1	42.5	54.8	72.9	90.0
	11	1	44.6	52.1	20.9	17.9	48.9	64.7	71.4	41.0	25.3	54.8	71.4	78.4	48.1	34.9	48.1	78.4	85.6	56.0	50.4	60.9	85.6	76.0	50.4	48.8	61.3	90.0
		12	43.4	64.7	34.9	10.6	34.9		78.4	41.0	20.6	41.0	71.4		47.3		51.9		85.6	56.0	46.6	46.6	78.4	90.0	60.9	46.6	90.0
Aug. 21	5	7	85.4	35.8	66.2	90.0	90.0	90.0	27.4	64.4	90.0	90.0	90.0	21.7	63.5	90.0	90.0	90.0	18.7	63.5	90.0	90.0	90.0	19.3	63.9	90.0	90.0	90.0
	6	6	79.0	25.4	51.8	87.9	90.0	90.0	15.8	49.8	90.0	90.0	90.0	7.3	49.2	77.8	90.0	90.0	7.4	50.1	90.0	90.0	90.0	12.3	51.3	82.3	90.0	90.0
	7	5	72.4	19.7	37.4	73.4	90.0	90.0	11.3	35.2	75.4	90.0	90.0	12.3	35.3	77.8	87.9	90.0	14.0	37.9	80.5	90.0	90.0	19.3	40.4	69.8	90.0	90.0
	8	4	66.1	22.6	23.0	58.8	78.6	90.0	20.2	20.7	60.8	83.5	90.0	22.3	22.8	73.6	87.9	90.0	22.9	24.3	80.5	88.6	90.0	32.2	30.5	58.2	90.0	90.0
	9	3	60.4	32.5	29.5	44.5	64.4	81.3	32.1	11.2	46.3	68.9	87.7	42.7	19.9	50.0	72.9	90.0	57.0	24.7	54.8	79.4	90.0	46.3	35.2	48.2	82.7	90.0
	10	2	55.8	43.5	21.2	29.5	50.0	68.6	46.2	25.1	32.2	54.2	74.0	57.0	29.4	37.2	59.5	81.8	71.7	39.5	44.6	65.4	90.0	60.8	35.2	41.0	64.5	90.0
	11	1	52.8	55.9	35.6	14.9	35.6	68.6	60.5	39.6	19.3	39.6	87.8	71.7	39.5	26.8	59.5	87.8	86.3	51.9	32.3	65.4	90.0	75.4	44.6	38.3	69.2	90.0
		12	51.7	68.6	35.6	2.3	35.6		74.2	45.2	12.3	45.2		86.3	51.9	22.3		86.3	86.3	51.9	32.3	56.3		90.0	56.3	56.3		90.0
Sept. 21	7	5	83.5	32.0	41.2	75.2	90.0	90.0	23.4	35.8	75.8	90.0	90.0	16.3	32.2	75.2	90.0	90.0	13.5	31.3	75.9	90.0	90.0	15.0	32.1	76.5	90.0	90.0
	8	4	77.3	32.9	28.2	60.5	84.9	90.0	29.0	12.2	60.0	84.8	90.0	26.8	17.3	60.5	87.9	90.0	27.8	18.6	62.0	90.0	90.0	30.0	21.5	63.3	90.0	90.0
	9	3	71.7	43.3	18.4	45.5	67.5	87.7	39.5	11.8	45.0	84.8	90.0	40.1	13.6	45.9	87.9	90.0	45.8	24.0	48.4	76.4	90.0	45.0	18.3	50.5	78.6	90.0
	10	2	67.7	51.1	17.4	31.5	55.0	87.1	52.0	25.6	30.0	72.9	90.0	54.2	13.2	31.5	72.9	90.0	57.5	20.2	35.5	76.4	90.0	60.0	25.3	38.9	78.6	90.0
	11	1	64.9	62.7	26.7	18.0	53.3	75.1	65.3	40.2	15.0	55.9	90.0	68.6	28.0	10.0	57.9	90.0	72.4	33.2	24.8	76.4	90.0	75.0	37.2	29.8	64.5	90.0
		12	64.0	75.0	39.5	10.0			78.9	40.2	12.3	39.6		83.1	43.0	22.3	43.0	78.9	87.4	47.3	20.0	51.9		90.0	50.5	26.0		90.0
Oct. 21	8	4	87.0	44.0	35.3	63.1	90.0	90.0	38.0	26.4	60.6	86.1	90.0	33.6	18.6	58.9	87.1	90.0	31.5	13.8	58.4	88.8	90.0	31.6	13.8	58.5	89.6	90.0
	9	3	81.9	49.2	27.4	36.2	84.9	90.0	43.0	19.7	46.6	71.5	90.0	45.1	12.4	45.5	72.7	90.0	45.1	2.8	44.3	74.3	90.0	46.0	8.1	45.2	75.5	90.0
	10	2	77.9	58.6	34.1	25.4	71.0	90.0	57.7	19.7	31.6	57.2	90.0	59.3	15.8	30.4	57.9	90.0	59.3	15.8	30.4	59.7	90.0	60.6	19.0	32.1	61.3	90.0
	11	1	75.4	69.2	44.9	20.5	57.5	90.0	70.2	43.1	10.5	43.1	83.3	73.8	30.5	0.5	43.2	85.8	73.8	30.5	9.5	45.1	88.4	75.3	32.9	21.0	47.0	90.0
		12	74.5	81.0	44.9		44.9		83.3	43.1	10.5	43.1		88.4	45.1	0.5	43.2		88.4	45.1	9.5	45.1		90.0	47.0	15.5	47.0	90.0
Nov. 21	10	2	87.0	65.7	36.5	42.0	74.6	90.0	63.5	28.1	35.4	73.5	90.0	62.2	21.1	30.5	72.9	90.0	61.7	17.1	28.2	72.9	90.0	61.9	17.2	28.3	73.1	90.0
	11	1	86.6	73.4	41.7	41.7	62.1	90.0	74.9	36.0	24.7	60.0	90.0	74.9	30.7	17.5	58.9	90.0	75.4	30.7	14.1	58.8	89.4	75.9	31.3	15.1	59.3	90.0
		12	83.8	86.4	50.8	29.8	50.8		87.3	47.2	19.8	47.2		88.3	44.7	9.8	45.1	88.3	89.4	44.7	0.2	44.7		90.0	45.3	6.2	45.3	90.0
Dec. 21	11	1	88.2	77.9	44.8	36.5	64.0	90.0	76.9	38.6	27.6	61.3	90.0	76.3	33.9	19.6	59.5	90.0	76.2	31.5	14.3	58.7	90.0	76.3	31.3	13.9	58.8	90.0
		12	87.4	88.5	53.3	33.4	53.3		88.9	49.1	23.4	49.1		89.3	46.3	13.5	46.3	89.3	89.7	45.0	3.4	45.0		90.0	45.1	2.5	45.1	90.0

aBased on data in Table 1, p. 387, 1972 *ASHRAE Handbook of Fundamentals;* 0% ground reflectance, 1.0 clearness factor.

Figure A.1. Mean Daily Insolation in Btu/day/ft^2 (reprinted from "HUD Intermediate Minimum Property Standards Supplement," 1977 edition).

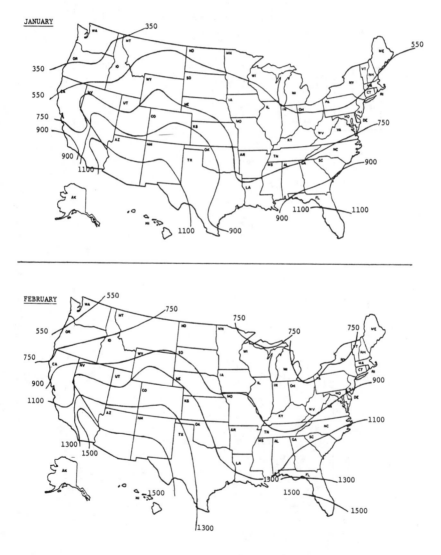

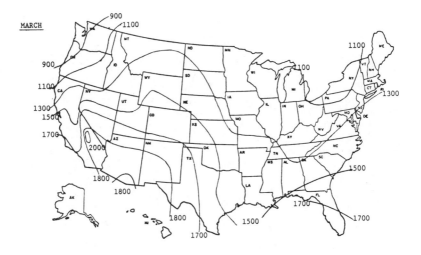

MARCH

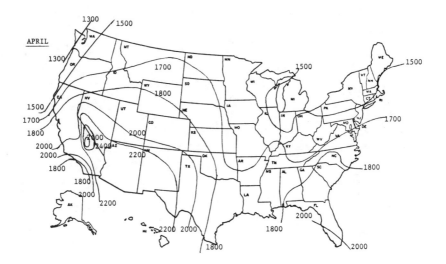

APRIL

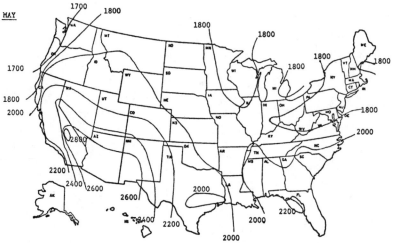

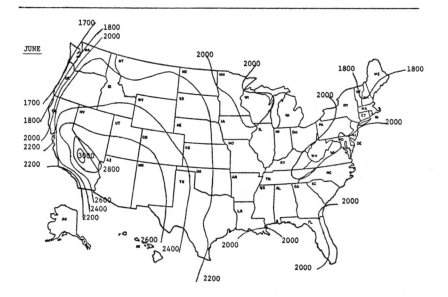

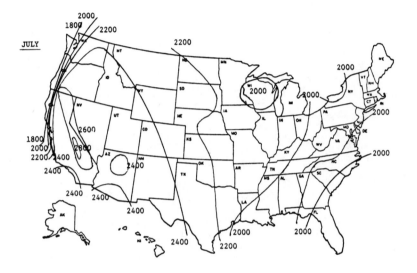

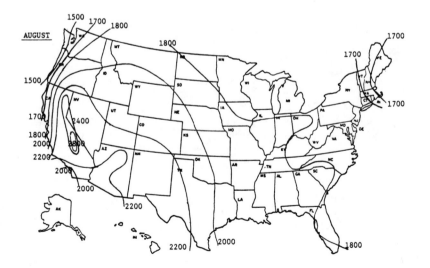

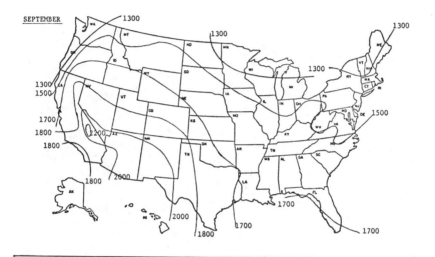

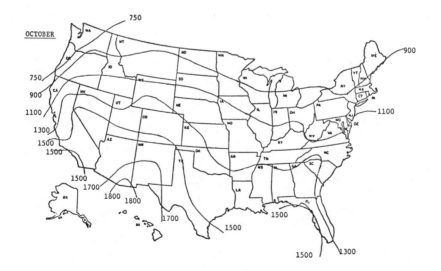

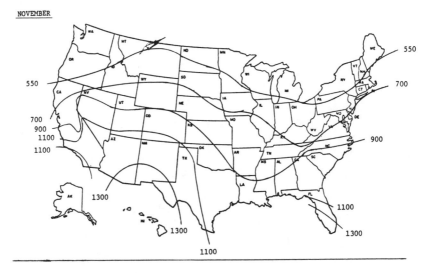

NOVEMBER

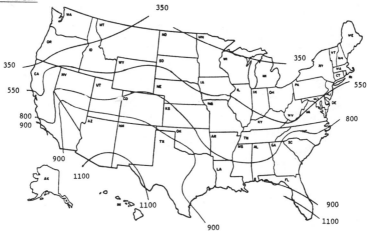

DECEMBER

Absorbent: of the two working fluids used in absorption cooling, this is the less volatile.

Alternating Current (ac): an electric current that reverses direction (most often 60 times per second), to allow for manipulation of voltage and current.

Ambient: characteristic of the immediate locality.

Biogas: anaerobic digestion results in the evolution of this substance, primarily composed of methane and carbon dioxide, which contains about half the heating value of natural gas.

Biomass: animal or plant tissue that comes into being as a result of biological growth or metabolism. As an input to power production, it refers specifically to plant tissue and animal excreta.

British Thermal Unit (Btu): a basic unit of energy equivalent to that energy necessary to raise the temperature of one pound of water one degree Fahrenheit.

Chlorophyll: a greenish chemical substance occurring in most plants that is instrumental to the plant's ability to carry out the photosynthetic process.

Conduction: one of the three modes of heat transfer, in which adjacent molecules in physically touching solid materials directly exchange thermal energy. The most efficient means of heat transfer.

Convection: one of the three modes of heat transfer, in which a fluid (gas or liquid) undergoes thermally induced changes in density, causing convective movement. The resulting physical movement of the fluid carries heat from one location to another.

Direct Current (dc): the major form of electric current in which the flow of electrons proceeds in one direction only.

Energy: the capacity to perform work. Energy exists in a myriad of forms, all of which theoretically can be converted into one another.

This convertibility is a basic and universal phenomenon, as are the inevitable losses encountered during any conversion process.

Fahrenheit: a scale of temperature measurement related to the behavior of water at atmospheric pressure, in which water freezes at approximately 32° and boils at 212°.

Heat: the total energy contained by randomly moving molecules in any material.

Heat Engine: a device that is able, by means of a working fluid, to convert a temperature differential existing between two regions into mechanical energy.

Heat Exchanger: a device, usually consisting of a coiled arrangement of metal tubing, that is used to transfer heat through the tube walls from one fluid to another.

Heat Pump: a device that converts input mechanical energy into a temperature differential increase by means of a working fluid. The device is used to pump heat from a cooler region to a warmer region, contrary to the direction of "natural" heat flow. May be assisted or augmented by solar thermal energy.

Heat Transfer Fluid: a liquid or gas that is used to absorb heat at some location and to surrender it after physically carrying it to some other location in a system.

Heliostat: the driving system used to orient a subarray of mirrors such that the mirror can reflect the sun's direct energy to a specific point, regardless of the sun's apparent above-horizon position.

Horsepower: a unit of power equal to 33,000 ft-lb/min.

Infrared Radiation: literally meaning "below the red," infrared energy possesses wavelengths of 0.8-7.0 μ, somewhat longer than that observable by the human eye. Sometimes called "heat rays," since they can be sensed as warmth by the skin.

Insolation: a term used to describe the intensity of sunlight (of all wavelengths) falling on a unit area. "Total insolation" includes the direct, diffuse and reflected components of sunlight at the terrestrial surface.

Insulation: refers to various materials that are used because of their ability to retard the flow of heat.

Kinetic Energy: the energy possessed by a mass (solid, liquid or gas) in motion. That energy must be absorbed to slow the mass down, a phenomenon that makes devices such as windmills and turbines possible.

Methane: a colorless and odorless gas whose molecule is composed of one atom of carbon and four atoms of hydrogen. Methane is the simplest hydrocarbon compound.

Micron: a very small unit of length equal to one-millionth of a meter. Current practice is beginning to replace this term with "micrometer."

Photon: the smallest possible unit of radiant energy. The energy content of a photon depends entirely on its wavelength. The shorter the wavelength, the greater the energy.

Photosynthesis: the process by which green plants manufacture food by using solar energy to combine carbon dioxide and water. Oxygen is a "waste" product of the process.

Power: the rate at which work is done, in units of energy per unit time.

Radiation: one of the three modes of heat transfer, in which radiant energy (capable of traveling through a vacuum) is given off by one body and intercepted at a distance by another. Examples of radiant energy include the entire solar spectrum.

Refrigerant: as used in absorption cooling cycles, the more volatile of the two working fluids. Capable of boiling at relatively low temperatures (much lower than water).

Solar Constant: the rate at which earth receives insolation from the sun, at a level just above the atmosphere. The rate undergoes very small variations throughout each year.

Specific Heat: the heat necessary to raise the temperature of a unit mass of a specific material by one degree. Each discrete material has its own characteristic specific heat, and those with the higher values are those most useful for purposes of heat storage.

Ultraviolet Radiation: the wavelengths of ultraviolet rays are too short to be visible to humans, having lengths of 0.008-0.4 μ. Popularly known as "black light," ultraviolet radiation is the most energetic component of that sunlight that reaches the terrestrial surface.

Work: that which is done when a force moves an object. Equal to the magnitude of the forces multiplied by the distance of the movement, work is a primary way in which to measure energy.

Working Fluid: a fluid used to transfer heat from one place in a system to another, often chosen for its ability to change phase at appropriate temperatures and pressures.

APPENDIX C

CONVERSION FACTORS

Multiply	By	To Obtain
Acres	43.560	Square feet
Acres	4,047	Square meters
Acres	1.562×10^{-3}	Square miles
Acres	4840	Square yards
Acre-feet	43.560	Cubic-feet
Acre-feet	3.259×10^{5}	Gallons
Angstrom units	3.937×10^{-9}	Inches
Atmospheres	76.0	Centimeters of mercury
Atmospheres	29.92	Inches of mercury
Atmospheres	33.90	Feet of water
Atmospheres	10,333	Kilograms/square meter
Atmospheres	14.70	Pounds/square inch
Atmospheres	1.058	Tons/square foot
Barrels (British, dry)	5.780	Cubic feet
Barrels (British, dry)	0.1637	Cubic meters
Barrels (British, dry)	36	Gallons (British)
Barrels, cement	170.6	Kilograms
Barrels, cement	376	Pounds of cement
Barrels, oil	42	Gallons (U.S.)
Barrels, (U.S., liquid)	4.211	Cubic feet
Barrels, (U.S., liquid)	0.1192	Cubic meters
Barrels (U.S., liquid)	31.5	Gallons (U.S.)
Bars	0.9869	Atmospheres
Bars	1×10^{6}	Dynes/square centimeter
Bars	1.020×10^{4}	Kilograms/square meter
Bars	2.089×10^{3}	Pounds/square foot
Bars	14.50	Pounds/square inch

Multiply	By	To Obtain
Board-feet	144 square inches x 1 inch	Cubic inches
British thermal units	0.2520	Kilogram-calories
British thermal units	777.5	Foot-pounds
British thermal units	3.927×10^{-4}	Horsepower-hours
British thermal units	1054	Joules
British thermal units	107.5	Kilogram-meters
British thermal units	2.928×10^{-4}	Kilowatt-hours
Btu (mean)	251.98	Calories, gram (mean)
Btu (mean)	0.55556	Centigrade heat units
Btu (mean)	6.876×10^{-5}	Pounds of carbon to CO_2
Btu/minute	12.96	Foot-pounds/second
Btu/minute	0.02356	Horsepower
Btu/minute	0.01757	Kilowatts
Btu/minute	17.57	Watts
Btu/square foot/ minute	0.1220	Watts/square inch
Btu (mean)/hour $(ft^2)°F$	4.882	Kilogram-calorie/ $(m^2)°C$
Btu (mean)/hour $(ft^2)°F$	1.3562×10^{-4}	Gram-calorie/second $(cm^2)°C$
Btu (mean)/hour $(ft^2)°F$	3.94×10^{-4}	Horsepower/$(ft^2)°F$
Btu (mean)/hour $(ft^2)°F$	5.682×10^{-4}	Watts/$(cm^2)°C$
Btu (mean)/hour $(ft^2)°F$	2.035×10^{-3}	Watts/$(in.^2)°C$
Btu (mean)/pound/ °F	1	Calories, gram/gram/°C
Bushels	1.244	Cubic feet
Bushels	2150	Cubic inches
Bushels	0.03524	Cubic meters
Bushels	4	Pecks
Bushels	64	Pints (dry)
Bushels	32	Quarts (dry)
Calories, gram (mean)	3.9685×10^{-3}	Btu (mean)
Calories, gram (mean)	0.001469	Cubic feet-atmospheres
Calories, gram (mean)	3.0874	Foot-pounds
Calories, gram (mean)	0.0011628	Watt-hours
Calories, (thermochemical)	0.999346	Calories (int. steam tables)

Multiply	By	To Obtain
Calories, gram (mean)/gram	1.8	Btu (mean)/pound
Centigrams	0.01	Grams
Centiliters	0.01	Liters
Centimeters	0.0328083	Feet (U.S.)
Centimeters	0.3937	Inches
Centimeters	0.01	Meters
Centimeters	393.7	Mils
Centimeters	10	Millimeters
Centimeter-dynes	1.020×10^{-3}	Centimeter-grams
Centimeter-dynes	1.020×10^{-8}	Meter-kilograms
Centimeter-dynes	7.376×10^{-8}	Pound-feet
Centimeter-grams	980.7	Centimeter-dynes
Centimeter-grams	10^{-5}	Meter-kilograms
Centimeter-grams	7.233×10^{-5}	Pound-feet
Centimeters of mercury	0.01316	Atmospheres
Centimeters of mercury	0.4461	Feet of water
Centimeters of mercury	136.0	Kilograms/square meter
Centimeters of mercury	27.85	Pounds/square foot
Centimeters of mercury	0.1934	Pounds/square inch
Centimeters/second	1.969	Feet/minute
Centimeters/second	0.03281	Feet/second
Centimeters/second	0.036	Kilometers/hour
Centimeters/second	0.6	Meters/minute
Centimeters/second	0.02237	Miles/hour
Centimeters/second	3.728×10^{-4}	Miles/minute
Centimeters/second/ second	0.03281	Feet/second/second
Centimeters/second/ second	0.036	Kilometers/hour/second
Centimeters/second/ second	0.02237	Miles/hour/second
Circular mils	5.067×10^{-6}	Square centimeters
Circular mils	7.854×10^{-7}	Square inches
Circular mils	0.7854	Square mils
Cord-feet	4 feet x 4 feet x 1 foot	Cubic feet

Multiply	By	To Obtain
Cords	8 feet x 4 feet x 4 feet	Cubic feet
Cubic centimeters	3.531×10^{-5}	Cubic feet
Cubic centimeters	6.102×10^{-2}	Cubic inches
Cubic centimeters	10^{-6}	Cubic meters
Cubic centimeters	1.308×10^{-6}	Cubic yards
Cubic centimeters	2.642×10^{-4}	Gallons
Cubic centimeters	10^{-3}	Liters
Cubic centimeters	2.113×10^{-3}	Pints (liquid)
Cubic centimeters	1.057×10^{-3}	Quarts (liquid)
Cubic centimeters	0.033814	Ounces (U.S. fluid)
Cubic feet	2.832×10^{4}	Cubic centimeters
Cubic feet	1728	Cubic inches
Cubic feet	0.02832	Cubic meters
Cubic feet	0.03704	Cubic yards
Cubic feet	7.481	Gallons
Cubic feet	28.32	Liters
Cubic feet	59.84	Pints (liquid)
Cubic feet	29.92	Quarts (liquid)
Cubic feet of water (60°F)	62.37	Pounds
Cubic feet/minute	472.0	Cubic centimeters/second
Cubic feet/minute	0.1247	Gallons/second
Cubic feet/minute	0.4720	Liters/second
Cubic feet/minute	62.4	Pounds of water/minute
Cubic feet/second	1.9834	Acre-feet/day
Cubic feet/second	448.83	Gallons/minute
Cubic feet/second	0.64632	Million gallons/day
Cubic feet-atmospheres	2.7203	Btu (mean)
Cubic foot-atmospheres	680.74	Calories, gram (mean)
Cubic foot-atmospheres	2116.3	Foot-pounds
Cubic foot-atmospheres	292.6	Kilogram-meters
Cubic foot-atmospheres	7.968×10^{-4}	Kilowatt-hours
Cubic inches	16.39	Cubic centimeters
Cubic inches	5.787×10^{-4}	Cubic feet
Cubic inches	1.639×10^{-5}	Cubic meters
Cubic inches	2.143×10^{-5}	Cubic yards
Cubic inches	4.329×10^{-3}	Gallons
Cubic inches	1.639×10^{-2}	Liters
Cubic inches	0.03463	Pints (liquid)
Cubic inches	0.01732	Quarts (liquid)
Cubic inches (U.S.)	0.55411	Ounces (U.S. fluid)

Multiply	By	To Obtain
Cubic meters	10^6	Cubic centimeters
Cubic meters	35.31	Cubic feet
Cubic meters	61,023	Cubic inches
Cubic meters	1.308	Cubic yards
Cubic meters	264.2	Gallons
Cubic meters	10^3	Liters
Cubic meters	2113	Pints (liquid)
Cubic meters	1057	Quarts (liquid)
Cubic meters	8.1074×10^{-4}	Acre-feet
Cubic meters	8.387	Barrels (U.S., liquid)
Cubic yards (British)	0.9999916	Cubic yards (U.S.)
Cubic yards	7.646×10^5	Cubic centimeters
Cubic yards	27	Cubic feet
Cubic yards	46.656	Cubic inches
Cubic yards	0.7646	Cubic meters
Cubic yards	202.0	Gallons
Cubic yards	764.6	Liters
Cubic yards	1616	Pints (liquid)
Cubic yards	807.9	Quarts (liquid)
Cubic yards/minute	0.45	Cubic feet/second
Cubic yards/minute	3.367	Gallons/second
Cubic yards/minute	12.74	Liters/second
Days	1440	Minutes
Days	86,400	Seconds
Decigrams	0.1	Grams
Deciliters	0.1	Liters
Decimeters	0.1	Meters
Degrees (angle)	60	Minutes
Degrees (angle)	0.01745	Radians
Degrees (angle)	3600	Seconds
Degrees/second	0.01745	Radians/second
Degrees/second	0.1667	Revolutions/minute
Degrees/second	0.002778	Revolutions/second
Dekagrams	10	Grams
Dekaliters	10	Liters
Dekameters	10	Meters
Drams	1.772	Grams
Drams	0.0625	Ounces
Dynes	1.020×10^{-3}	Grams
Dynes	7.233×10^{-5}	Poundals
Dynes	2.248×10^{-6}	Pounds

Multiply	By	To Obtain
Dynes per square centi-meter	1	Bars
Ergs	9.486×10^{-11}	British thermal units
Ergs	1	Dyne-centimeters
Ergs	7.376×10^{-8}	Foot-pounds
Ergs	1.020×10^{-3}	Gram-centimeters
Ergs	10^{-7}	Joules
Ergs	2.390×10^{-11}	Kilogram-calories
Ergs	1.020×10^{-8}	Kilogram-meters
Ergs/second	5.692×10^{-9}	British thermal units/minute
Ergs/second	4.426×10^{-6}	Foot-pounds/minute
Ergs/second	7.376×10^{-8}	Foot-pounds/second
Ergs/second	1.341×10^{-10}	Horsepower
Ergs/second	1.434×10^{-9}	Kilogram-calories/minute
Ergs/second	10^{-10}	Kilowatts
Fathoms	6	Feet
Feet	30.48	Centimeters
Feet	12	Inches
Feet	0.3048	Meters
Feet	1/3	Yards
Feet (U.S.)	1.893939×10^{-4}	Miles (statute)
Feet of air (1 atmos-phere 60°F)	5.30×10^{-4}	Pounds/square inch
Feet of water	0.02950	Atmospheres
Feet of water	0.8826	Inches of mercury
Feet of water	304.8	Kilograms/square meter
Feet of water	62.43	Pounds/square foot
Feet of water	0.4335	Pounds/square inch
Feet/minute	0.5080	Centimeters/second
Feet/minute	0.01667	Feet/second
Feet/minute	0.01829	Kilometers/hour
Feet/minute	0.3048	Meters/minute
Feet/minute	0.01136	Miles/hour
Feet/second	30.48	Centimeters/second
Feet/second	1.097	Kilometers/hour
Feet/second	0.5921	Knots/hour
Feet/second	18.29	Meters/minute
Feet/second	0.6818	Miles/hour
Feet/second	0.01136	Miles/minute

Multiply	By	To Obtain
Feet/100 feet	1	Percent Grade
Feet/second/second	30.48	Centimeters/second/second
Feet/second/second	1.097	Kilometers/hour/second
Feet/second/second	0.3048	Meters/second/second
Feet/second/second	0.6818	Miles/hour/second
Foot-poundals	3.9951×10^{-5}	Btu (mean)
Foot-poundals	0.0421420	Joules (abs)
Foot-pounds	0.013381	Liter-atmospheres
Foot-pounds	3.7662×10^{-4}	Watt-hours (abs)
Foot-pounds	1.286×10^{-3}	British thermal units
Foot-pounds	1.356×10^{7}	Ergs
Foot-pounds	5.050×10^{-7}	Horsepower-hours
Foot-pounds	1.356	Joules
Foot-pounds	3.241×10^{-4}	Kilogram-calories
Foot-pounds	0.1383	Kilogram-meters
Foot-pounds	3.766×10^{-7}	Kilowatt-hours
Foot-pounds/minute	1.286×10^{-3}	British thermal units/minute
Foot-pounds/minute	0.01667	Foot-pounds/second
Foot-pounds/minute	3.030×10^{-5}	Horsepower
Foot-pounds/minute	3.241×10^{-4}	Kilogram-calories/minute
Foot-pounds/minute	2.260×10^{-5}	Kilowatts
Foot-pounds/second	7.717×10^{-2}	British thermal units/minute
Foot-pounds/second	1.818×10^{-3}	Horsepower
Foot-pounds/second	1.945×10^{-2}	Kilogram-calories/minute
Foot-pounds/second	1.356×10^{-3}	Kilowatts
Foot-pounds/second	4.6275	Btu (mean)/hour
Foot-pounds/second	1.35582	Watts (abs)
Gallons (British)	4516.086	Cubic centimeters
Gallons (British)	1.20094	Gallons (KU
Gallons (British)	10	Pounds (avordupois) of of water at 62°F
Gallons (U.S.)	128	Ounces (U.S. fluid)
Gallons	3785	Cubic centimeters
Gallons	0.1337	Cubic feet
Gallons	231	Cubic inches
Gallons	3.785×10^{-3}	Cubic meters
Gallons	4.951×10^{-3}	Cubic yards
Gallons	3.785	Liters

Multiply	By	To Obtain
Gallons	8	Pints (liquid)
Gallons	4	Quarts (liquid)
Gallons/minute	2.228×10^{-3}	Cubic feet/second
Gallons/minute	0.06308	Liters/second
Grains (troy)	1	Grains (average)
Grains (troy)	0.06480	Grams
Grains (troy)	0.04167	Pennyweights (troy)
Grains (troy)	2.0833×10^{-3}	Ounces (troy)
Grains/U.S. gallons	17.118	Parts/million
Grains/U.S. gallons	142.86	Pounds/million gallons
Grains/Imperial gallons	14.286	Parts/million
Grams	980.7	Dynes
Grams	15.43	Grains (troy)
Grams	10^{-3}	Kilograms
Grams	10^{3}	Milligrams
Grams	0.03527	Ounces
Grams	0.03215	Ounces (troy)
Grams	0.07093	Poundals
Grams	2.205×10^{-3}	Pounds
Gram-calories	3.968×10^{-3}	British thermal units
Gram-centimeters	9.302×10^{-8}	British thermal units
Gram-centimeters	980.7	Ergs
Gram-centimeters	7.233×10^{-5}	Foot-pounds
Gram-centimeters	9.807×10^{-5}	Joules
Gram-centimeters	2.344×10^{-8}	Kilogram-calories
Gram-centimeters	10^{-5}	Kilogram-meters
Gram-centimeters	2.7241×10^{-8}	Watt-hours
Gram-centimeters/ second	9.80665×10^{-5}	Watts (abs)
Grams-centimeters² (moment of inertia)	3.4172×10^{-4}	Pounds-inch²
Grams-centimeters²	2.37305×10^{-6}	Pounds-feet²
Grams/cubic meters	0.43700	Grains/cubic foot
Grams/centimeter	5.600×10^{-3}	Pounds/inch
Grams/cubic centimeter	62.43	Pounds/cubic foot
Grams/cubic centimeter	0.03613	Pounds/cubic inch
Grams/cubic centimeter	3.405×10^{-7}	Pounds/mil foot
Grams/cubic centimeter	8.34	Pounds/gallon
Grams/liter	58.417	Grains/gallon (U.S.)
Grams/liter	9.99973×10^{-4}	Grams/cubic centimeter
Grams/liter	1000	Parts/million (ppm)

Multiply	By	To Obtain
Grams/liter	0.06243	Pounds/cubic foot
Grams/square centi-meter	0.0142234	Pounds/square inch
Hectograms	100	Grams
Hectoliters	100	Liters
Hectometers	100	Meters
Hectowatts	100	Watts
Hemispheres (sol. angle)	0.5	Sphere
Hemispheres (sol. angle)	4	Spherical right angles
Hemispheres (sol. angle)	6.283	Steradians
Horsepower	42.44	British thermal units/minute
Horsepower	33,000	Foot-pounds/minute
Horsepower	550	Foot-pounds/second
Horsepower	1.014	Horsepower (metric)
Horsepower	10.70	Kilogram calories/minute
Horsepower	0.7457	Kilowatts
Horsepower	745.7	Watts
Horsepower (boiler)	33,520	British thermal units/hour
Horsepower (boiler)	9.804	Kilowatts
Horsepower, electrical	1.0004	Horsepower
Horsepower (metric)	0.98632	Horsepower
Horsepower-hours	2547	British thermal units
Horsepower-hours	1.98×10^6	Foot-pounds
Horsepower-hours	2.684×10^6	Joules
Horsepower-hours	641.7	Kilogram-calories
Horsepower-hours	2.737×10^5	Kilogram-meters
Horsepower-hours	0.7457	Kilowatt-hours
Hours	60	Minutes
Hours	3600	Seconds
Inches	2.540	Centimeters
Inches	10^3	Mils
Inches of mercury	0.03342	Atmospheres
Inches of mercury	1.133	Feet of water
Inches of mercury	0.0345	Kilograms/square centi-meters
Inches of mercury	345.3	Kilograms/square meter
Inches of mercury	25.40	Millimeters of mercury

Multiply	By	To Obtain
Inches of mercury	70.73	Pounds/square foot
Inches of mercury	0.4912	Pounds/square inch
Inches of water	0.002458	Atmospheres
Inches of water	0.07355	Inches of mercury
Inches of water	25.40	Kilograms/square meter
Inches of water	0.5781	Ounces/square inch
Inches of water	5.204	Pounds/square foot
Inches of water	0.03613	Pounds/square inch
Kilograms	980,665	Dynes
Kilograms	10^3	Grams
Kilograms	70.93	Poundals
Kilograms	2.2046	Pounds
Kilograms	1.102×10^{-3}	Tons (short)
Kilogram-calories	3.968	British thermal units
Kilogram-calories	3086	Foot-pounds
Kilogram-calories	1.558×10^{-3}	Horsepower-hours
Kilogram-calories	426.6	Kilogram-meters
Kilogram-calories	1.162×10^{-3}	Kilowatt-hours
Kilogram-calories/ minute	51.43	Foot-pounds/second
Kilogram-calories/ minute	0.09351	Horsepower
Kilogram-calories/ minute	0.06972	Kilowatts
Kilogram-centimeters^2	2.373×10^{-3}	Pounds-feet^2
Kilogram-centimeters^2	0.3417	Pounds-inches^2
Kilogram-meters	9.302×10^{-3}	British thermal units
Kilogram-meters	9.807×10^7	Ergs
Kilogram-meters	7.233	Foot-pounds
Kilogram-meters	3.6529×10^{-6}	Horsepower-hours
Kilogram-meters	9.579×10^{-6}	Pounds water evaporated at 212°F
Kilogram-meters	9.807	Joules
Kilogram-meters	2.344×10^{-3}	Kilogram-calories
Kilogram-meters	2.724×10^{-6}	Kilowatt-hours
Kilograms/cubic meter	10^{-3}	Grams/cubic meter
Kilograms/cubic meter	0.06243	Pounds/cubic foot
Kilograms/cubic meter	3.613×10^{-5}	Pounds/cubic inch
Kilograms/cubic meter	3.405×10^{-10}	Pounds/mil foot
Kilograms/meter	0.6720	Pounds/foot

Multiply	By	To Obtain
Kilograms/square centi-meter	28.96	Inches of mercury
Kilograms/square centi-meter	735.56	Millimeters of mercury
Kilograms/square centi-meter	14.22	Pounds/square inch
Kilograms/square meter	9.678×10^{-5}	Atmospheres
Kilograms/square meter	3.281×10^{-3}	Feet of water
Kilograms/square meter	2.896×10^{-3}	Inches of mercury
Kilograms/square meter	0.07356	Millimeters of mercury at $0°C$
Kilograms/square meter	0.2048	Pounds/square foot
Kilograms/square meter	1.422×10^{-3}	Pounds/square inch
Kilograms/square millimeter	10^6	Kilograms/square meter
Kiloliters	10^3	Liters
Kilometers	10^5	Centimeters
Kilometers	3281	Feet
Kilometers	10^3	Meters
Kilometers	0.6214	Miles
Kilometers	1093.6	Yards
Kilometers/hour	27.78	Centimeters/second
Kilometers/hour	54.68	Feet/minute
Kilometers/hour	0.9113	Feet/second
Kilometers/hour	0.5396	Knots/hour
Kilometers/hour	16.67	Meters/minute
Kilometers/hour	0.6214	Miles/hour
Kilometers/hour/ second	27.78	Centimeters/second/ second
Kilometers/hour/ second	0.9113	Feet/second/second
Kilometers/hour/ second	0.2778	Meters/second/second
Kilometers/hour/ second	0.6214	Miles/hour/second
Kilometers/minute	60	Kilometers/hour
Kilowatts	56.92	British thermal units/ minute
Kilowatts	4.425×10^4	Foot-pounds/minute
Kilowatts	737.6	Foot-pounds/second
Kilowatts	1.341	Horsepower

Multiply	By	To Obtain
Kilowatts	14.34	Kilogram-calories/ minute
Kilowatts	10^3	Watts
Kilowatt-hours	3415	British thermal units
Kilowatt-hours	2.655×10^6	Foot-pounds
Kilowatt-hours	1.341	Horsepower hours
Liters	10^3	Cubic centimeters
Liters	0.03531	Cubic feet
Liters	61.02	Cubic inches
Liters	10^{-3}	Cubic meters
Liters	1.308×10^{-3}	Cubic yards
Liters	0.2642	Gallons
Liters	2.113	Pints (liquid)
Liters	1.057	Quarts (liquid)
Liters/minute	5.885×10^{-4}	Cubic feet/second
Liters/minute	4.403×10^{-3}	Gallons/second
$Log_{10}N$	2.303	$Log_E N$ or Ln N
Log N or Ln N	0.4343	$Log_{10}N$
Meters	100	Centimeters
Meters	3.2808	Feet
Meters	39.37	Inches
Meters	10^{-3}	Kilometers
Meters	10^3	Millimeters
Meters	1.0936	Yards
Meters	10^{10}	Angstrom units
Meters	6.2137×10^4	Miles
Meter-kilograms	9.807×10^7	Centimeter-dynes
Meter-kilograms	10^5	Centimeter-grams
Meter-kilograms	7.233	Pound-feet
Meters/minute	1.667	Centimeters/second
Meters/minute	3.281	Feet/minute
Meters/minute	0.05468	Feet/second
Meters/minute	0.06	Kilometers/hour
Meters/minute	0.03728	Miles/hour
Meters/second	196.8	Feet/minute
Meters/second	3.281	Feet/second
Meters/second	3.6	Kilometers/hour
Meters/second	0.06	Kilometers/minute
Meters/second	2.237	Miles/hour
Meters/second	0.03728	Miles/minute
Meters/second/second	3.281	Feet/second/second

Multiply	By	To Obtain
Meters/second/second	3.6	Kilometers/hour/second
Meters/second/second	2.237	Miles/hour/second
Micrograms	10^{-6}	Grams
Microliters	10^{-6}	Liters
Microns	10^{-6}	Meters
Miles	1.609×10^5	Centimeters
Miles	5280	Feet
Miles	1.6093	Kilometers
Miles	1760	Yards
Miles (int. Nautical)	1.852	Kilometers
Miles/hour	44.70	Centimeters/second
Miles/hour	88	Feet/minute
Miles/hour	1.467	Feet/second
Miles/hour	1.6093	Kilometers/hour
Miles/hour	26.82	Meters/minute
Miles/hour/second	44.70	Centimeters/second/second
Miles/hour/second	1.467	Feet/second/second
Miles/hour/second	1.6093	Kilometers/hour/second
Miles/hour/second	0.4470	Meters/second/second
Miles/minute	2682	Centimeters/second
Miles/minute	88	Feet/second
Miles/minute	1.6093	Kilometers/minute
Miles/minute	60	Miles/hour
Milliers	10^3	Kilograms
Milligrams	10^{-3}	Grams
Milliliters	10^{-3}	Liters
Millimeters	0.1	Centimeters
Millimeters	0.03937	Inches
Millimeters	39.37	Mils
Millimeters of mercury	0.0394	Inches of mercury
Millimeters of mercury	1.3595^{-3}	Kilograms/square centimeter
Millimeters of mercury	0.01934	Pounds/square inch
Mils	0.002540	Centimeters
Mils	10^{-3}	Inches
Mils	25.40	Microns
Minutes (angle)	2.909×10^{-4}	Radians
Minutes (angle)	60	Seconds (angle)
Months	30.42	Days
Months	730	Hours

Multiply	By	To Obtain
Months	43,800	Minutes
Months	2.628×10^6	Seconds
Myriagrams	10	Kilograms
Myriameters	10	Kilometers
Myriawatts	10	Kilowatts
Ounces	16	Drams
Ounces	437.5	Grains
Ounces	28.35	Grams
Ounces	0.0625	Pounds
Ounces (fluid)	1.805	Cubic inches
Ounces (fluid)	0.02957	Liters
Ounces (U.S. fluid)	29.5737	Cubic centimeters
Ounces (U.S. fluid)	1/128	Gallons (U.S.)
Ounces (troy)	480	Grains (troy)
Ounces (troy)	31.10	Grams
Ounces (troy)	20	Pennyweights (troy)
Ounces (troy)	0.08333	Pounds (troy)
Ounces/square inch	0.0625	Pounds/square inch
Parts/million	0.0584	Grains/U.S. gallon
Parts/million	0.7016	Grains/Imperial gallon
Parts/million	8.345	Pounds/million gallons
Pennyweights (troy)	24	Grains (troy)
Pennyweights (troy)	1.555	Grams
Pennyweights (troy)	0.05	Ounces (troy)
Pints (dry)	33.60	Cubic inches
Pints (liquid)	28.87	Cubic centimeters
Pints (U.S. liquid)	473.179	Cubic centimeters
Pints (U.S. liquid)	16	Ounces (U.S. fluid)
Poundals	13,826	Dynes
Poundals	14.10	Grams
Poundals	0.03108	Pounds
Pounds	444,823	Dynes
Pounds	7000	Grains
Pounds	453.6	Grams
Pounds	16	Ounces
Pounds	32.17	Poundals
Pound (troy)	0.8229	Pounds (av.)
Pounds (troy)	373.2418	Grams
Pounds of carbon to CO_2	14,544	Britith thermal units (mean)
Pound-feet (torque)	1.3558×10^7	Dyne-centimeters

Multiply	By	To Obtain
Pound-feet	1.356×10^7	Centimeters-dynes
Pound-feet	13,825	Centimeter-grams
Pound-feet	0.1383	Meter-kilograms
Pounds-feet^2	421.3	Kilogram-centimeters^2
Pounds-feet^2	144	Pounds-inches^2
Pounds-inches^2	2,926	Kilogram-centimeters^2
Pounds-inches^2	6.945×10^{-3}	Pounds-feet^2
Pounds of water	0.01602	Cubic feet
Pounds of water	27.68	Cubic inches
Pounds of water	0.1198	Gallons
Pounds of water evaporated at 212°F	970.3	British thermal units
Pounds of water/ minute	2.699×10^{-4}	Cubic feet/second
Pounds/cubic foot	0.01602	Grams/cubic centimeter
Pounds/cubic foot	16.02	Kilograms/cubic meter
Pounds/cubic foot	5.787×10^{-4}	Pounds/cubic inch
Pounds/cubic foot	5.456×10^{-9}	Pounds/mil foot
Pounds/cubic inch	27.68	Grams/cubic centimeter
Pounds/cubic inch	2.768×10^4	Kilograms/cubic meter
Pounds/cubic inch	1728	Pounds/cubic foot
Pounds/cubic inch	9.425×10^{-6}	Pounds/mil foot
Pounds/foot	1.488	Kilograms/meter
Pounds/inch	178.6	Grams/centimeter
Pounds/square foot	0.01602	Feet of water
Pounds/square foot	4.882	Kilograms/square meter
Pounds/square foot	6.944×10^{-3}	Pounds/square inch
Pounds/square inch	0.06804	Atmospheres
Pounds/square inch	2.307	Feet of water
Pounds/square inch	2.036	Inches of mercury
Pounds/square inch	0.0703	Kilograms/square centimeter
Pounds/square inch	703.1	Kilograms/square meter
Pounds/square inch	144	Pounds/square foot
Pounds/square inch	70.307	Grams/square centimeter
Pounds/square inch	51.715	Millimeters of mercury at 0°C
Quadrants (angle)	90	Degrees
Quadrants (angle)	5400	Minutes
Quadrants (angle)	1.571	Radians·
Quarts (dry)	67.20	Cubic inches

Multiply	By	To Obtain
Quarts (liquid)	57.75	Cubic inches
Quarts (U.S. liquid)	0.033420	Cubic feet
Quarts (U.S. liquid)	32	Ounces (U.S. fluid)
Quarts (U.S. liquid)	0.832674	Quarts (British)
Radians	57.30	Degrees
Radians	3438	Minutes
Radians	0.637	Quadrants
Radians/second	57.30	Degrees/second
Radians/second	0.1592	Revolutions/second
Radians/second	9.549	Revolutions/minute
Radians/second/second	573.0	Revolutions/minute/ minute
Radians/second/second	9.549	Revolutions/minute/ second
Radians/second/second	0.1592	Revolutions/second/ second
Revolutions	360	Degrees
Revolutions	4	Quadrants
Revolutions	6.283	Radians
Revolutions/minute	6	Degrees/second
Revolutions/minute	0.1047	Radians/second
Revolutions/minute	0.01667	Revolutions/second
Revolutions/minute/ minute	1.745×10^{-3}	Radians/second/second
Revolutions/minute/ minute	0.01667	Revolutions/minute/ second
Revolutions/minute/ minute	2.778×10^{-4}	Revolutions/second/ second
Revolutions/second	360	Degrees/second
Revolutions/second	6.283	Radians/second
Revolutions/second	60	Revolutions/minute
Revolutions/second/ second	6.283	Radians/second/second
Revolutions/second/ second	3600	Revolutions/minute/ minute
Revolutions/second/ second	60	Revolutions/minute/ minute
Seconds (angle)	4.848×10^{-6}	Radians
Spheres (solid angle)	12.57	Steradians
Spherical right angles	0.25	Hemispheres
Spherical right angles	0.125	Spheres

Multiply	By	To Obtain
Spherical right angles	1.571	Steradians
Square centimeters	1.973×10^5	Circular mils
Square centimeters	1.076×10^{-3}	Square feet
Square centimeters	0.1550	Square inches
Square centimeters	10^{-6}	Square meters
Square centimeters	100	Square millimeters
Square centimeters-centimeters squared	0.02420	Square inches-inches squared
Square feet	2.296×10^{-5}	Acres
Square feet	929.0	Square centimeters
Square feet	144	Square inches
Square feet	0.09290	Square meters
Square feet	3.587×10^{-8}	Square miles
Square feet	1/9	Square yards
Square feet-feet squared	2.074×10^4	Square inches-inches squared
Square inches	1.273×10^6	Circular mils
Square inches	6.452	Square centimeters
Square inches	6.944×10^{-3}	Square feet
Square inches	10^6	Square mils
Square inches	645.2	Square millimeters
Square inches (U.S.)	7.71605×10^{-4}	Square yards
Square inches-inches squared	41.62	Square centimeters-centimeters squared
Square kilometers	247.1	Acres
Square kilometers	10.76×10^6	Square feet
Square kilometers	10^6	Square meters
Square kilometers	0.3861	Square miles
Square kilometers	1.196×10^6	Square yards
Square meters	2.471×10^{-4}	Acres
Square meters	10.764	Square feet
Square meters	3.861×10^{-7}	Square miles
Square meters	1.196	Square yards
Square miles	640	Acres
Square miles	27.88×10^6	Square feet
Square miles	2.590	Square kilometers
Square miles	3.098×10^6	Square yards
Square millimeters	1.973×10^3	Circular mils
Square millimeters	0.01	Square centimeters
Square millimeters	1.550×10^{-3}	Square inches
Square mils	1.273	Circular mils

Multiply	By	To Obtain
Square mils	6.452×10^{-6}	Square centimeters
Square mils	10^{-6}	Square inches
Square yards	2.066×10^{-4}	Acres
Square yards	9	Square feet
Square yards	0.8361	Square meters
Square yards	3.228×10^{-7}	Square miles
Temperature (°C) + 273	1	Absolute temperature (°C)
Temperature (°C) + 17.8	1.8	Temperature (°F)
Temperature (°F) + 460	1	Absolute temperature (°F)
Temperature (°F) - 32	5/9	Temperature (°C)
Tons (long)	1016	Kilograms
Tons (long)	2240	Pounds
Tons (metric)	10^3	Kilograms
Tons (metric)	2205	Pounds
Tons (short)	907.2	Kilograms
Tons (short)	2000	Pounds
Tons (short)/square feet	9765	Kilograms/square meter
Tons (short)/square feet	13.89	Pounds/square inch
Tons (short)/square inch	1.406×10^6	Kilograms/square meter
Tons (short)/square inch	2000	Pounds/square inch
Watts	0.05692	British thermal units/minute
Watts	10^7	Ergs/second
Watts	44.26	Foot-pounds/minute
Watts	0.7376	Foot-pounds/second
Watts	1.341×10^{-3}	Horsepower
Watts	0.01434	Kilogram-calories/minute
Watts	10^{-3}	Kilowatts
Watt-hours	3.415	British thermal units
Watt-hours	2655	Foot-pounds
Watt-hours	1.341×10^{-3}	Horsepower-hours
Watt-hours	0.8605	Kilogram-calories
Watt-hours	367.1	Kilogram-meters
Watt-hours	10^{-3}	Kilowatt-hours
Weeks	168	Hours
Weeks	10,080	Minutes
Weeks	604,800	Seconds

Multiply	By	To Obtain
Yards	91.44	Centimeters
Yards	3	Feet
Yards	36	Inches
Yards	0.9144	Meters
Years (common)	365	Days
Years (common)	8760	Hours

REFERENCES

1. Meinel, A. B., and M. P. Meinel. *Applied Solar Energy—An Introduction* (Reading, MA: Addison-Wesley Publishing Co., 1976).
2. Rau, H. *Solar Energy* D. J. Duffin, Ed. (New York: The Macmillan Company, 1964).
3. Gabel, M. *Earth, Energy and Everyone* (California: Straight Arrow Books, 1975).
4. Williams, J. R. *Solar Energy—Technology and Applications* (Ann Arbor, MI: Ann Arbor Science Publishers, Inc., 1974).
5. "Burning Question," *Scientific Am.* 236 (6):64 (1977).
6. Bronowski, J. *The Ascent of Man* (Boston: Little, Brown and Company, 1973).
7. Halacy, D. S., Jr. *Fabulous Fireball—The Story of Solar Energy* (New York: The Macmillan Company, 1957).
8. Daniels, F. *Direct Use of the Sun's Energy* (New Haven, CT: Yale University Press, 1964).
9. Jordan, R. C., and W. E. Ibele. "Mechanical Energy From Solar Energy," in *Proceedings of the World Symposium of Applied Solar Energy* (Menlo Park: Stanford Research Institute, 1965).
10. Herwig, L. O. "U.S. Overview of Solar Energy," in *Japanese/United States Symposium on Solar Energy Systems,* Vol. 2 (Washington, DC: The MITRE Corporation, 1974).
11. Szokolay, S. V. *Solar Energy and Building* (New York: John Wiley and Sons, Inc., 1975).
12. Anderson, B. *The Solar Home Book* (Harrisville: Cheshire Books, 1976).
13. *ASHRAE Applications Handbook* (New York: American Society of Heating, Refrigeration and Air Conditioning Engineers, 1974).
14. Duffie, J. A., and W. A. Beckman. *Solar Energy Thermal Processes* (New York: John Wiley and Sons, Inc., 1974).
15. Brinkworth, B. J. *Solar Energy For Man* (London: The Compton Press Ltd., 1972).
16. Schumacher, E. F. *An Economics of Permanance,* Institute for the Study of Non-Violence.
17. "Patterns of Energy Consumption in the United States," *Stanford Research Institute Report* (January 1972).
18. Marcovich, S. J. "Autonomous Living," *Popular Sci.* 207 (6):82 (1975).

19. Tabor H. "Selective Surfaces for Solar Collectors," in *Applications of Solar Energy for Heating and Cooling of Buildings* (New York: American Society of Heating, Refrigeration and Air Conditioning Engineers, 1977).

20. Stepler, R. "Now You Can Buy Solar Heating Equipment for Your Home," *Popular Sci.* 206(3):77 (1975).

21. Close, D. J. "Solar Air Heaters for Low and Moderate Temperature Applications," *J. Solar Energy Sci. Eng.* 7(3) (July 1963).

22. Britton, P. "World's Most Advanced Solar Home," *Popular Sci.* 211 (1):94 (1977).

23. Hoff, J.E. "Sunpak Solar Energy Collector Background Information," Owens-Illinois Corp. (1977).

24. "The Owens-Illinois Sunpak Solar Collector," Owens-Illinois brochure No. 5-2877-24-10M (1975).

25. "The Hot-Line Solar Collector," *The Mother Earth News* No. 39, 108 (1976).

26. Gilmore, C. P. "Concentrating Collectors for Solar Heating and Cooling," *Popular Sci.* 209 (4):97 (1976).

27. Leckie, J., G. Masters, H. Whitehouse and L. Young. *Other Homes and Garbage,* Sierra Club Books (1975).

28. *Solar Energy Q's and A's,* Sunworks (1976).

29. "Direct Uses of Solar Energy," in *Energy for Rural Development— Renewable Resources and Alternative Technologies for Developing Countries* (Washington, DC: National Academy of Sciences, 1976).

30. Löf, G. O. G. "Systems for Space Heating with Solar Energy," in *Applications of Solar Energy for Heating and Cooling of Buildings* (New York: American Society of Heating, Refrigeration and Air Conditioning Engineers, 1977).

31. Barber, E., and D. Watson. *Design Criteria for Solar Heated Buildings,* Sunworks Technical Publication (1975).

32. *Design and Construction of a Residential Solar Heating and Cooling System,* prepared by the Solar Energy Applications Laboratory of Colorado State University for the National Science Foundation (Springfield, VA: National Technical Information Service, 1974).

33. Daniels, G. *Solar Homes and Sun Heating* (New York: Harper & Row Publishers, 1976).

34. Newton, A. B. "Solar Cooling," *Applications of Solar Energy for Heating and Cooling of Buildings* (New York: American Society of Heating, Refrigeration and Air Conditioning Engineers, 1977).

35. Wilbur, P. J., and C. E. Mitchell. "Solar Absorption Air Conditioning Alternatives," *Solar Energy* (3):193-199 (1975).

36. Brannon, P. J., *et al. The Environmental Issues Associated with Solar Heating and Cooling of Residential Dwellings*, Sandia Laboratories, Albuquerque, New Mexico (April 1977).

37. Morse, R. N., and D. J. Close. "Solar Water Heating," in *Applications of Solar Energy for Heating and Cooling of Buildings* (New York: American Society of Heating, Refrigeration and Air Conditioning Engineers, 1977).

38. Cohen, S., Ed. "Solar Energy....World View," *Consulting Eng.* 48 (3):104-110 (1977).

39. Sloggett, G. "Energy Used for Pumping Irrigation Water in the United States, 1974, *Solar Irrigation Workshop Proceedings,* Sandia Laboratories, Albuquerque, New Mexico (1977).

40. "Solar Powers Irrigation Pump," *Solar Eng. Mag.* 2 (8):26-28 (1977).

41. Albis, R. L., and J. M. Alcone. *Solar Powered Irrigation System,* Sandia Laboratories, Albuquerque, New Mexico (September 1976).

42. *A Solar Energy Experiment—Solar Powered Irrigation,* Sandia Laboratories, Albuquerque, New Mexico.

43. Murphy, L. M., and A. C. Skinrood. *Development of the Solar Power Central Receiver Concept,* Sandia Laboratories, Albuquerque, New Mexico (September 1976).

44. Bradley, W. J. "Designing and Siting Solar Power Plants," *Consulting Eng.* 48 (3)80-84 (1977).

45. Backus, C. E., Ed., *Solar Cells* (New York: Institute of Electrical and Electronics Engineers (IEEE) Press, 1976).

46. Chalmers, B. "The Photovoltaic Generation of Electricity," *Scientific Am.* 235 (4):34-43 (1976).

47. Bickler, D. B., and E. N. Costoque. "Photovoltaic Cells and Arrays," in *Record of the Photovoltaics Power Conditioning Workshop,* Sandia Laboratories, Albuquerque, New Mexico (March 1977).

48. Angrist, S. W. *Direct Energy Conversion,* 2nd ed. (Boston, MA: Allyn and Bacon, Inc., 1971).

49. Hickok, F. *Handbook of Solar and Wind Energy* (Boston, MA: Cahners Books, 1975).

50. Currin, C. G., *et al.* "Feasibility of Low Cost Silicon Solar Cells," *Solar Cells* (New York: IEEE Press, 1976).

51. Dermatis, S. N., and J. W. Faust, Jr. "Semiconductor Sheets for the Manufacture of Semiconductor Devices," *IEEE Transactions Communications Electronics* (New York: IEEE Press, 1963).

52. Robinson, A. L. "Amorphous Silicon: A New Direction for Semiconductors," *Science* 197(4306):851-853 (1977).

53. Buch, F., A. L. Fahrenbruch and R. H. Bube. "Photovoltaic Properties of Five II-VI Heterojunctions," *J. Appl. Phys.* 48(4):1596-1602 (1977).

54. Hammond, A. L. Photovoltaics: The Semiconductor Revolution Comes to Solar," *Science,* 197(4302):445-447 (1977).

55. Beckman, W. A., *et al.* "Design Considerations for a 50 Watt Photovoltaic Power System Using Concentrated Solar Energy," *Solar Energy* 10(3) (1966).

56. Raunels, J. E. "Solar Total Energy Program," *Solar Total Energy Symposium Proceedings* Sandia Laboratories: Albuquerque, New Mexico (January 1977).

57. Tarui, Y. "Japanese Photovoltaic Systems," *Japanese/United States Symposium on Solar Energy Systems* (Washington, DC: The MITRE Corporation, 1974).

58. *NASA Report to Educators* 5(3) (October 1977).

59. *MITRE Photovoltaic Energy Systems* (Washington, DC: The MITRE Corporation, 1975).

60. *In Situ Performance Measurements of the MITRE Photovoltaic Array* (Washington, DC: The MITRE Corporation, 1976).

61. "Irrigation Calls for Cell Power," *Solar Eng. Mag.* 2(11):28-29 (1977).
62. "Photovoltaic Energy Program Summary," *Solar Energy Program Anthology* (Washington, DC: Energy Research and Development Administration (ERDA), 1977).
63. Grey, J., P. Downey and B. Davis, Eds. "New Potentials—Space Based Solar Power," in *Space—A Resource For Earth* (New York: American Institute of Aeronautics and Astronautics (AIAA), 1977).
64. Glaser, P. E. "Development of the Satellite Solar Power Station," in *Space Manufacturing Facilities—Space Colonies* (New York: AIAA, Inc., 1977).
65. Summers, R. A., H. R. Blieden, and C. E. Bloomquist. "Assessment of Satellite Power Stations," *Third Princeton/AIAA Conference on Space Manufacturing Facilities* (New York: AIAA, 1977).
66. O'Neill, G. K. *The High Frontier—Human Colonies in Space* (New York: William Morrow and Company, Inc., 1977)
67. Halacy, D. S., Jr. *Earth, Water, Wind, and Sun—Our Energy Alternatives* (New York: Harper and Row Publishers, 1977).
68. Betz, A. "Windmills in the Light of Modern Research," [U.S. National Advisory Committee for Aeronautics, *Technical Memorandum No. 474* (August 1928)] from *Die Naturwiss.* XV(46) (November 1927).
69. Sencenbaugh, J. "Wind Driven Generators," *Energy Primer—Solar, Water, Wind, and Biofuels,* Portola Institute (Fremont, CA: Fricke-Parks Press, 1974).
70. Ragsdale, L., and R. E. Christensen. "City of the Sun Previewed in Model," *Consulting Eng.* 48(3)98-100 (1977).
71. Putnam, P. C. *Power From the Wind* (New York: Van Nostrand Reinhold Company, 1948).
72. Torrey, V. *Wind Catchers—American Windmills of Yesterday and Tomorrow* (Brattleboro, VT: The Stephen Greene Press, 1976).
73. Lindsley, E. F. "Wind Power: How New Technology is Harnessing an Age-Old Energy Source," *Popular Sci.* 205(1)54-59; 124-125 (1974).
74. Simmons, D. M. *Wind Power* (Park Ridge, NJ: Noyes Data Corp., 1975).
75. Kocivar, B. "World's Largest Windmill—Turns On for Large-Scale Wind Power," *Popular Sci.* 208(3):73-75; 150-151 (1976).
76. Golding, E. W. *The Generation of Electricity by Wind Power,* 3rd ed. (London: E. & F.N. Spon Ltd., 1977).
77. Hackleman, M. A. *Wind and Windspinners* (Culver City, CA: Peace Press, 1974).
78. Shepard, M. L., *et al. Introduction to Energy Technology* (Ann Arbor, MI: Ann Arbor Science Publishers, Inc., 1976).
79. Prenis, J., Ed. *Energy Book 2* (Philadelphia, PA: Running Press, 1977).
80. Sforza, P. M. "Vortex Augmentor Concepts for Wind Energy Conversion," *ERDA and NSF Wind Energy Workshop* (New York: Polytechnic Institute of New York, 1975).
81. Loth, J. L. "Wind Energy Concentrators," in *Proceedings of the Second Annual UMR-MEC Conference on Energy* (Hollywood, CA: Western Periodicals, 1976).
82. Kociva, B. "Tornado Turbine Reaps Power from Whirlwind," *Popular Sci.* 210(1):78-80 (1977).

83. Wilkerson, A. W. *Synchronous Inversion Techniques for Utilization of Waste Energy* (Mukwonago, WI: Windworks, 1976).

84. Eldridge, F. "Wind Energy Conversion Systems with Compressed-Air Storage," *Preliminary Projections for Centralized Usage of Solar Energy Systems in 1980, 1985 and 1990* (McLean, VA: MITRE Corporation, 1976).

85. Meyer, H. *Synchronous Inversion-Concepts and Applications* (Mukwonago, WI: Windworks 1976).

86. Crawley, G. M. *Energy* (New York: Macmillan Publishing Co., Inc., 1975).

87. Heroneumus, W. E. "Toward a National Energy Policy," *Chancellor's Lecture Series,* University of Massachusetts, Amherst, April 21, 1977.

88. Trimble, L. C. "Engineering Aspects of OTEC Systems," presented at the Spring Meeting/STAR Symposium of the Society of Naval Architects and Marine Engineer, San Francisco, California, May 25-27, 1977.

89. Goodenough, J. B. "The Options for Using the Sun," *Technol. Rev.* 79(1):63-71 (1976).

90. Beorse, B. *OTEC History,* University of California (September 1977).

91. Dugger, G. L., E. L. Francis and W. H. Avery. "Technical and Economic Feasibility of Ocean Thermal Energy Conversion," presented at Sharing the Sun! Solar Technology in the Seventies—a Joint Conference of the American Section ISES and the SESC, Winnipeg, Manitoba, Canada, August 15-20, 1976.

92. Griffin, O. U. "Power From the Oceans' Thermal Gradients," *Sea Technol.* 18(8):11-15; 38-40 (1977).

93. *Ocean Thermal Energy Conversion,* Lockheed Missiles and Space Company, Inc., California.

94. *Ocean Thermal Energy Conversion—Final Report,* Volume 4 of Test Program Plant, TRW Systems Group, (California, June 1975).

95. Zener, C. "The OTEC Answer to OPEC-Solar Sea Power!" *Mech. Eng.* V99, N. 6, 26-29 (June 1977).

96. Tschupp, E., B. Berkowitz, and W. Hausz. *OTEC Implementation Problems for Specific Missions,* TEMPO—Center for Advanced Studies, General Electric Company, Washington, DC.

97. Naef, F. E., "Economic Aspects of Ocean Thermal Energy Conversion," Lockheed Missiles and Space Company, Inc., presented at Sharing the Sun! Solar Technology in the Seventies—a Joint Conference of the American Section ISES and the SESC, Winnipeg, Manitoba, Canada, August 15-20, 1976.

98. Hornburg, C. D., *et al.* "Preliminary Research on Ocean Energy Industrial Complexes," *Proceedings of Joint ISES/SESC Conference, Sharing the Sun! Solar Energy in the Seventies,* Winnipeg, Manitoba, Canada, August 15-20, 1976.

99. Dugger, G. L. Private communication (February 1978).

100. "Proposed Research Program on Continued Evaluation of the Commercial Feasibility of Hydrogen Production by the Solar Photoelectrolysis of Water," Battelle Columbus Laboratories, October 1977, prepared for the U.S. Department of Energy.

101. Wrighton, M. S. "The Chemical Conversion of Sunlight," *Technol. Rev.* 79 (6):30-37 (May 1977).

102. Bockris, J. O. *Energy: The Solar Hydrogen Alternative* (New York: Halstead Press, Div. of John Wiley & Sons, 1975).
103. Nozik, A. J. *Energy Conversion Via Photoelectrolysis*, Materials Research Center, Allied Chemical Corporation, Morristown, NJ.
104. Nozik, A. J. "Photoelectrolysis of Water Using Semiconducting TiO_2 Crystals," *Nature* 257(5525):383-386 (1975).
105. Nozik, A. J. "Electrode Materials for Photoelectrochemical Devices," *J. of Crystal Growth* 39:200-209 (1977).
106. Chang, K. C., *et al.* "Stable Semiconductor Liquid Junction Cell With 9 Percent Solar-to-Electrical Conversion Efficiency," *Science* 196:1097-1099 (1977).
107. Miller, B., and A. Heller. "Semiconductor Liquid Junction Solar Cells Based on Anodic Sulfide Films," *Nature* 262(5570):680-681 (1976).
108. Miller, B., *et al.* "Solar Conversion Efficiency of Pressure Sintered Cadmium Selenide Liquid Junction Cells," *J. Electrochem. Soc.* 124 (7):1019-1021 (1977).
109. Heller, A., K. C. Chang, and B. Miller. "Spectral Response and Efficiency Relations in Semiconductor Liquid Junction Solar Cells," *J. Electrochem. Soc.* 124(5):697-700 (1977).
110. Stoker, H. S., S. L. Seage, and R. L. Capener. *Energy From Source to Use* (Glenview, IL: Scott, Foresman & Company, 1975).
111. Ramakumar, R. "An Assessment of Hydrogen as a Means to Store Solar Energy," *Proceedings of Joint ISES/SESC Conference, Sharing the Sun. Solar Energy in the Seventies*, Winnipeg, Manitoba, Canada, August 15-20, 1976.
112. Hsu, M., S.S., W. E. Morrow, and J. B. Goodenough. "High Efficiency Electrochemical Plant," *Record of the Tenth Intersociety Energy Conversion Engineering Conference*, University of Delaware, Newark, Delaware, August 18-22, 1975.
113. Bockris, J. O., and D. M. Drazic. *Electro-Chemical Science* (New York: Barnes and Noble Books, 1972).
114. Gates, D. M. "The Flow of Energy in the Biosphere," *Energy and Power* (San Francisco, CA: W. H. Freeman and Company, 1971).
115. Merrill, R. "Biomass Energy," *Energy Primer—Solar, Water, Wind, and Biofuels*, Portola Institute (Fremont, CA: Fricke-Park Press, 1974).
116. *The Renewable Biomass Energy Guidebook*, The Biomass Energy Institute, Inc., Winnipeg, Manitoba, Canada (1974).
117. Alich, J. A., Jr., and R. E. Inman. "Energy from Agriculture," *Record of the Tenth Intersociety Energy Conversion Engineering Conference*, University of Delaware, Newark, Delaware, August 18-22, 1975.
118. Inman, R. E. "Silviculture Energy Plantations," *Proceedings of Joint ISES/SESC Conference, Sharing the Sun! Solar Energy in the Seventies*, Winnipeg, Manitoba, Canada, August 15-20, 1976.
119. Lipinsky, E. S. "Field Crops as a Future Source of Fuels and Chemical Feedstocks," *Proceedings of Joint ISES/SESC Conference, Sharing the Sun! Solar Energy, in the Seventies*, Winnipeg, Canada, August 15-20, 1976.
120. Cornwell, D. A., *et al.* "Nutrient Removal by Water Hyacinths," *J. Water Poll. Control Fed.*

121. Fischer, A. W., Jr. "Engineering for Algae Culture," *Proceedings of the World Symposium on Applied Solar Energy,* Stanford Research Institute, Menlo Park, CA (1956).

122. Hammond, A. L. "Photosynthetic Solar Energy: Rediscovering Biomass Fuels," *Science* 197(4305):745-746 (1977).

123. Council on Environmental Quality. "Energy and the Environment Electric Power" (August 1973),p. 40

124. Schellenbach, S., W. Turnacliff, and F. Varani. *Methane on the Move: A Discussion of Small Anaerobic Digesters,* Bio-Gas of Colorado, Inc. and Colorado Energy Research Institute (March 1977).

125. Schulte, D. D., *et al.* "Methane Production Through Bioconversion of Agricultural Residues," *Proceedings of Joint ISES/SESC Conference, Sharing the Sun! Solar Energy in the Seventies,* Winnipeg, Manitoba, Canada, August 15-20, 1976.

126. "Manure is Now a Commodity," *CALF News* 15(10):44-45 (1977).

127. Varani, F., J. Burford, and R. P. Arber. *The Design of a Large-Scale Manure/Methane Facility,* Bio-Gas of Colorado, Inc., and CH2M Hill, Inc. (June 1977).

128. *Resource Recycling: The Energy Potential of Bio-Conversion in the Four Corners Region,* Bio-Gas of Colorado, Inc., and Colorado Energy Research Institute (October 1976).

129. Schmid, L. W. "Feedlot Wastes to Useful Energy—Fact or Fiction?" *J. Environ. Eng. Div. ASCE* 101(EE5):787-793 (1975).

130. Cheremisinoff, P. N., and A. C. Morresi. *Energy From Solid Wastes* (New York: Marcel Dekker Inc., 1976).

INDEX

Abbot   13
absorber
   efficiency   40
   /fluid passage designs   39
   plate   35,46
   surface   34,36,46
absorptance   40,41
absorption cooling cycle   64
acceptance angle   51
activation energy   146
active systems   57
Adams   8
Advanced Concepts Division   118
   Flight Projects Laboratory   108
agriculture industry   69
air-cooled
   collectors   34,44,45,47,62
   heating systems   60
alcohol   153
Alternate Energy Resources Co.
   52
aluminum production   136
Amenkotep III   1
American Society of Heating, Refrigeration, and Air Conditioning Engineers (ASHRAE)
   26
ammonia   127,129,133,138,141,
   153
   cooling cycle   66
   production   137
amorphous (uncrystallized) silicon
   89
anaerobic digestion   163,166,167,
   170
angle of incidence   123
animal
   feedlots   164
   wastes   164

antifreeze solution   35,63
antireflective coating   84
Archimedes   1,3
Argonne National Laboratories
   48,55
array systems   92
auxiliary heat inputs   67
Averani   3
Avondale Shipyards   132
axicon-type concentration devices   9

bacterial degradation   164
band gap   146,148
Bari   16
Barr   9
baseboard heaters   45,61
Battelle Memorial Institute   70
Bechtel Corp.   126
Bell Telephone Laboratories   82
Bessemer   5
biofouling   125
Bio-Gas of Colorado, Inc.   167
biological conversion   157
biomass   158,159
   photosynthetic   159
   to energy   165
   yielding crops   162
borosilicate glass   49
Boyle   9
Boys   11
Buffon   3
busbar cost   131

cadmium
   selenide   90
   sulfide   82
   sulfide solar cell   89
   telluride   90

Callier   5
capital energy sources   32
Carnegie-Mellon University   125,
    127,134
casing requirement   46
Cassini   4
cell configuration   148
cellulose   159
C.F. Braun Co.   127
chemical
    conversion   141
    fuels   147
    potential energy   141,142
    process   141
    storage   121
chemotrophs   158
chlorella pyredenoisa   163
chlorophyll   158
closed-cycle turbine   124
collection
    element   58
    field layout   73
    surface   25
collector
    orientation   27
    performance   25
    plates   35
    Also see concentrating and
        nonconcentrating collectors
combustion   141,165,170
compact fuel   152
concentrating collectors   51
concentration ratio   51
conceptual design   76
conductive heat   101
conductors   81
controls   63
convective heat   101
conversion   34
cooling
    of buildings   64
    systems   60,66,67
corn   160,168
Corning Glass Works   48
cosmic radiation   20
cost   44,49,55,86,89,133,137
    effectiveness   41
    material   40
Cottle   8
crops   159

crystal growth   86
Czochralski
    cells   86
    crystals   85,87
    silicon solar cells   86

Darrieus rotor   114,115
desalt water   125
de Saussure   4
distribution   58
domestic
    hot water (DHW)   67
    water heating   33,67
DSS Engineers, Inc.   137
ducting systems   64
Dunn   8

Earthmind   114
economic factors   45
efficiency   86
EFG crystal growth   88
Elecktro GMBH Co.   114
electric current   81
electrical
    conversion   124,142
    generation   81,92,105
    insulators   81
    potential   84
    power production   166
electricity
    by cable transmission   135
    generation   71
electrochemical
    cell   142
    energy converter   154
    photovoltaic cells,   150,151
electrode
    instability   146,151
    materials   150
    stability   148
electrolysis   142
electrolyte cells   142
electrolytic cell   144
electromagnetic radiation   18
emittance   40,41
Eneas   9
    axicon concentrator   12

energy
    consumption   33
    gap   82
    plantation   162
    prices   32
    sources   32
    storage   141
    transmission   75
    Also see activation, solar ther-
        mal, sunlight-to-thermal,
        thermal and wind
enzymes   168
equator   22
Ericsson   6,7
ethanol   153
evacuated-tube collector   48,50,
    67
evaporated (working) fluid   124
evaporation effects   65
evaporator   132

fabrication
    cost   40
    technology   39,45
Federal Power Commission   107
fermentation   168
fiberglass   63
    insulation   53
Finplank
    absorber   38
    assembly   38
fixed mirror   71
flash-evaporating   124
flat
    black paints   41
    plate collector   10,33,34,35,
        43,58,67
    panels   35
fluid leakage   46
food   159
fossil fuels   141,153
Freon   113
    refrigerant   125
Fresnel lens   92
fuel
    crop   161
    output   152
Fujishima   142,145

Galen   1
gallium
    -arsenide   90
    -phosphide   90
gasoline   153,164
General Electric Co.   48,105
geographical latitudes   26
Ghai   15
glass   42
    Also see borosilicate glass, low
        iron glass
glazing   34,41,46,49
    a collector   43
Global Marine Development Inc.
    127
Goddard   13
Goldstone Tracking Station   98
greenhouse effect   42
Grumman   38,108,118

Harrington   13
Harvey   13
    collection efficiency   35
heat
    engine   129
    engine cycle 123
    exchanger   62
    intensity   33
    Also see auxiliary heat inputs,
        conductive heat, convective
        heat
heat transfer   34,46,63
    arrangements   38
    fluid   35,63,71
    surface   36
heating
    of buildings   58
    systems   58
heliostat   75
    field layout   78
Herschel   5
heterojunction cells   90
Heynemann   8
Himalaya   10
    solar furnace   11
Honda   142,145
horizontal
    -axis machines   103,108
    tube evaporators   132

Hot-Line
  collector   52
  concentrator   53
Hottel   14
house of the future   68
housing design   31
hydraulic pressure loss   40
hydro-storage   105
hydride   153
hydrogen   141,142,143,144,151, 153
  -burning automobile   153
  economy   155
  gas carriers   152
  -oxygen fuel cell   154
  storage   152
  utilization   152
hydrogasification   169
hydrogenation   169
hydronic systems   61
hydrozone   153

incident angle   26,27
income energy sources   32
indium-phosphate   90
insolation   24,34,49,84,101,142
  levels   45
  on inclined surfaces   175
  Also see mean daily insolation maps, solar insolation, terrestrial insolation.
installation costs   44
insulation   34,43,46
integral fluid passage-type absorber   38,39
iron titanium hydride   152
irrigation   70,73
  system   74

Japan   93
Johns Hopkins University   131, 138

Kaiser Aluminum & Chemical Corp.   132
kelp   163
Ketcham   9
Khanna   15
kinetic energy   102

Kircher   2

Langley   8
Lavoisier   4
  solar furnace   4
lead storage batteries   95
Lewis Research Center   93,110
Lincoln Laboratories   95
liquid-cooled
  collection systems   36
  flat-plate collectors 34,35,61
lithium bromide absorption   66
Lockheed   126,127,132
  OTEC design 129
Löf   15
low-iron glass   42

magnesium nickel hydride   152
manure   164
mariculture   139
Maritime Administration   132
McHenry   9
mean daily insolation maps   175
mechanical failure   40
Mennon   1
metallic hydride compounds   152
methane   153,167
methanol   153
microalgae   163
microwave
  radiation   98,100
  transmitter   98
mirror   75
  type collectors   10
M.I.T. Solar House   14,15
  I   34
MITRE Corp.   93
  Solar Energy System   94
Mobile Tyco Laboratories   86
Model-O   113
Molero   14
motion of the earth   22
Mouchot   5
  solar concentrator   6

NASA
  Goldstone Tracking Station 99
  Jet propulsion Lab   93

NASA
Langley Research Center 68
Venus antenna site 98
National Science Foundation 125
New Alchemy Institute 108
nonconcentrating collectors 33,
48
Northrup 53
concentrator 54
Nozik 147
N.V. Philips Gloeilampenfabrieken
48

ocean-thermal energy conversion
OTEC 124,144
conversion 164
design 130
gradient power 123
plant-ship 133
power applications 135,136
system 124
oceanic surface layer 123
Olin Brass Co. 39
optical transmission 75
Ouroboros House 39
Owens-Illinois 48
oxidation 142
reduction 144
oxygen-suppression experiments
149
ozone layer 19,20

parabolic
cylindrical reflectors 56
trough concentrators 71,73
passive systems 57
performance 41
photoanode 147
photoelectrochemical devices
142,151,155
photoelectrolysis 142,145,149,
153
cells 143,146-148,150
photoelectrolytic cells 145
photoexcitation 146,147
photosynthesis 139,157
photosynthetic plants 157
phototrophs 157,163

photovoltaic 70,90,141
arrays 92
cell 150
devices 84,151
effect 82,145
generation 81,95
systems 93
photovoltage 142
Pifre 6
pilot plant 76
planar reflective surfaces 51
plastics 42
platinum
counterelectrode 142
foil cathode 147
Pliny 2
Plutarch 1
p-n junction 82,146
photovoltaic cell 83
Poillet 3
polycrystalline
materials 84
semiconductors 151
potential energy 141
preheat tank 69
pressure solar engines 9
pressurized piping 36
Priestly 4
Princeton University 108
Project Sunshine 93
protective
casing 34
outer casing 43
protein 160
pumping applications 74
pyrolysis 169

radio-frequency communications
100
Rankine-cycle 71
RCA 82
Reagan 8
reflective surfaces 51
refraction (lenses) 51
refrigerant 65,66,71
refuse-derived fuel 165
regeneration 65
Revere Copper and Brass 39
ribbon crystals 86

Roll-Bond panels   38
Romagnoli   13
Royal Mirror   4

Sailwing   108
Salomon de Caux   2
Salter's rotor   111
Sandia Laboratories   71,75,91,116
    photovoltaic concentrator sys-
        tem   92
sandwich-type absorber   39
Savonius rotor   114,115,118
Schottky
    barrier   90
    -type cell   150
seasonal solar energy   155
selective surfaces   41
selenium   90
semiconductor   82,148,150
    electrolyte   141,150
    materials   82,88,145
    Also see polycrystalline materials
Severy   8
Shipman   13
Shuman   11
silicon   82,90
    crystals   82
    dioxide   84
    p-n junction cell   84
    ribbon growing   87
    solar cell   84,90
    wafers   84
    Also see amorphous silicon,
        Czochralski silicon solar cells
silviculture   162
single crystal electrode   151
sludge   166,167
Smith-Putnam wind turbine   103,
    105,106,111
sodium acetate   147
solar
    cell   82
    central receiver   79
    concentration   6
    DHW system   70
    engine   2
    flux   91
    furnace   4
    heated water   71

solar
    insolation   142
    irrigation   11,69,73,74
    oven   5
    position   21,23
    radiation   18,24,25
    satellite power station   96,97
    solid-state   151
    thermal energy   92
Solaris collectors   36
solid wastes   164
sorghum   160,161,168
sounding statues   1
Southwestern Engineering Co.
    127
space
    -based generation   95
    conditioning system   68
    heating   33,46,58,59,64
SSPS concept   96,98
stagnation temperatures   42
starchy grains   168
stationary concentration   52
Stock   8
storage
    element   58
    methodology   153
    tank   63
    volume   63
    water, heated   64
strontium titanate   147
submarine cable transmission
    136
sugar beets   160
sugarcane   160
Sun Shipbuilding & Dry Dock Co.
    132
sunlight-to-thermal energy   34
Sunpak Solar Collector   48
sunshine intensity   86
suntracking   71
Sweeney's Sailwind wind turbine
    109
swimming pool heating   35
synchronous generator   105

Targioni   3
terrestrial
    generation   92